THINKING NATURE *and the* NATURE OF THINKING

Cultural Memory

in

the

Present

Hent de Vries, Editor

THINKING NATURE AND
THE NATURE OF THINKING

From Eriugena to Emerson

Willemien Otten

STANFORD UNIVERSITY PRESS

Stanford, California

STANFORD UNIVERSITY PRESS
Stanford, California

Printed in the United States of America on acid-free, archival-quality paper

Library of Congress Cataloging-in-Publication Data

Names: Otten, Willemien, author.
Title: Thinking nature and the nature of thinking : from Eriugena to Emerson /
    Willemien Otten.
Description: Stanford, California : Stanford University Press, 2020. |
    Includes bibliographical references and index.
Identifiers: LCCN 2019028970 (print) | LCCN 2019028971 (ebook) |
    ISBN 9781503606708 (cloth) | ISBN 9781503611672 (paperback) |
    ISBN 9781503611689 (ebook)
Subjects: LCSH: Philosophy of nature. | Nature—Religious aspects. |
    Erigena, Johannes Scotus, approximately 810-approximately 877. |
    Emerson, Ralph Waldo, 1803–1882. | Natural theology.
Classification: LCC BD581 .O76 2020 (print) | LCC BD581 (ebook) |
    DDC 113—dc23
LC record available at https://lccn.loc.gov/2019028970
LC ebook record available at https://lccn.loc.gov/2019028971

Cover image: Kim Lim, from *Three Small Etchings*, "Untitled," 1971.
© Estate of Kim Lim. All Rights Reserved, DACS, London /
ARS, NY 2019; Photo: © Tate, London 2019

Cover design: Rob Ehle

Typeset by Kevin Barrett Kane in 11/13.5 Adobe Garamond Pro

For Dick

Contents

Abbreviations

CCCM	Corpus Christianorum Continuatio Mediaevalis
CCSL	Corpus Christianorum Series Latina
Conf.	*Confessions of Augustine*
CW	*Collected Works of Ralph Waldo Emerson*
Gen. litt.	*De Genesi ad litteram libri duodecim (On the literal meaning of Genesis) of Augustine*
Hom.	*Homilies* in *Hexaemeron of Basil of Caesarea*
KGA	*Kritische Gesamtausgabe* [of the works of F. D. Schleiermacher]
PG	*Patrologia Graeca*, edited by J. P. Migne
PL	*Patrologia Latina*, edited by J. P. Migne
Retract.	*Retractations of Augustine*

Acknowledgments

This book is the product of many years of reflection on John the Scot Eriugena and Ralph Waldo Emerson. I am grateful to my readers, Karmen McKendrick and Brian Stock, for their generous judgment in evaluating the manuscript, to series editor Hent de Vries for his unceasing faith in this project, and to my editor at Stanford University Press, Emily-Jane Cohen, for her supportive advice throughout the process.

For the last decade I have made my intellectual home at the University of Chicago in Hyde Park, which is where this book has come to fruition. In the Divinity School I have enjoyed the freedom to develop courses around sweeping ideas and have been surprised every quarter by the critical engagement of my students. Two seminars in particular have helped me shape this book: Creation East and West: From Origen to Eriugena, in the autumn of 2012, and The Religious Thought of R. W. Emerson and William James, in the autumn of 2017. I am grateful to the students in these seminars for their creative responses.

Aside from the fact that Eriugena is front and center in it, this book touches only lightly on medieval materials. I am nevertheless grateful to my colleagues—students and faculty—in the medieval studies community at the University of Chicago. The workshop meetings at noon on Fridays are a mainstay of our joint intellectual life, and I hope that the workshop continues to thrive.

My nomination as a 2015–16 Henry Luce III Fellow in Theology by the Association of Theological Schools in the United States and Canada (ATS) and the Henry Luce Foundation has been a major boon for this project, then called *Natura Educans: The Psychology of Pantheism*. In addition to thanking my recommenders, Burcht Pranger, David Tracy, and Hent de Vries, I am grateful to Michael Gilligan and Jonathan VanAntwerpen of the Luce Foundation for hosting the fellows at the Luce headquarters in New York, and to Stephen Graham and Leah Wright at ATS for hosting us at three successive fellows conferences in Pittsburgh. I thank James Wetzel (Villanova University) for serving as my interlocutor at one of these.

I also thank Kathleen Moore, chair of Religious Studies, and Thomas Carlson, my host, for inviting me to be the 2017 Tipton Visiting Distinguished Professor of Catholic Studies at the University of California, Santa Barbara. One advantage of spending the winter in Santa Barbara was my exposure to its magnificent scenery, another the opportunity to teach a course titled "Creation between God and Nothing: What Religious Authors Have to Say on Pantheism and Nature Mysticism." The colloquium I organized there, "Thinking Nature and the Nature of Thinking," with talks by Burcht Pranger (University of Amsterdam), Charles Hallisey (Harvard University), and Thomas Carlson, yielded the title of this book. My Tipton public lecture "Theological and Environmental Criticism: Creation, Pantheism, and the Problem of Thinking Nature" has appeared under the title "Nature as a Theological Problem: An Emersonian Response to Lynn White," in *Responsibility and the Enhancement of Life: Essays in Honor of William Schweiker*, edited by G. Thomas and H. Springhart (Leipzig: Evangelische Verlagsanstalt, 2017), 265–80.

The material in this book has not been published before except for parts of Chapter 2 and of the Postscript to Part 1. In Chapter 2, on Maximus the Confessor, I have made use of my articles "Cosmos and Liturgy from Maximus to Hans Urs von Balthasar (with an excursion on H. J. Schulz)," in *Sanctifying Texts, Transforming Rituals: Encounters in Liturgical Studies*, edited by P. van Geest, M. Poorthuis, and H. E. G. Rose (Leiden: Brill, 2017), 153–69; and "West and East: Prayer and Cosmos in Augustine and Maximus the Confessor," in *Prayer and the Transformation of the Self in Early Christian Mystagogy*, edited by H. van Loon, G. de Nie, M. Op de Coul, and P. van Egmond (Leuven: Peeters, 2018), 319–37; while the Postscript to Part 1 contains elements from the mystagogy article and from my earlier publication "Eriugena and Emerson on Nature and the Self," in *Eriugena and Creation*, edited by M. I. Allen and W. Otten (Turnhout: Brepols, 2014), 503–38. I am grateful to Brill, Peeters, and Brepols for their permission to make use of these materials.

I thank my successive deans for overseeing our collective work in the Divinity School. In my current position as director of the Martin Marty Center for the Public Understanding of Religion I greatly enjoy the support of

Julia Woods. It has been exciting to interact in this role with Martin Marty, Dean Laurie Zoloth, and Dean David Nirenberg, as we attempt to cultivate the informed discussion about religion through *Sightings* columns, the fellows seminar, public events, conferences, and a film festival. The advice of colleagues Yousef Casewit and Ryan Coyne on the conference "Political Theology: Promise and Prospects" has been much appreciated.

I am humbled by the scholarship of my graduate students, from whom I feel I am only beginning to learn. I owe a special debt to Elena Lloyd-Sidle, who has not only inspired my seminars on theological criticism but whose editorial skills have vastly improved the book's clarity of expression. Susan Schreiner, William Schweiker, Michael Allen, Jean-Luc Marion, Bruce Lincoln, Charlie Hallisey, and Babette Hellemans remind me that collegiality and friendship can be organic, dare I say natural. Bernie and Pat McGinn are not only friends but inexhaustible sources of warmth and wisdom. My longtime friend Burcht Pranger is a keen reader and precise thinker, whose advice and insight I continue to find invaluable.

All these years my family has been spread out over two continents and remains remarkably intact and tight-knit. As our daughters, Maud and Fu Cheng, find their own way in life, I dedicate this book to Derk Jansen, my spouse. He knows that I am deeply grateful but will appreciate that what cannot be put into words is better left unsaid.

THINKING NATURE *and the* NATURE OF THINKING

Thinking Nature . . . (and the Nature of Thinking)

Man carries the world in his head, the whole astronomy and chemistry suspended in a thought. Because the history of nature is charactered in his brain, therefore is he the prophet and discoverer of her secrets.

—Ralph Waldo Emerson, "Nature"

"Thinking nature" is the rubric under which this book brings together the religious thought of Johannes Scottus Eriugena (810–77) and Ralph Waldo Emerson (1803–82) with a number of other congenial minds, creating a Platonic symposium of sorts that stretches across the Western religious tradition. The aim of my project is to explore nature or creation as driving the structure of thought rather than being driven by it, which I think is the case in all of the thinkers treated here, despite their differences. I refer to this as their practice of "thinking nature" in order to bring out the human encounter with *nature in thought* rather than any instrumentalization or domestication of nature as the object of thought. The thinkers I have chosen form a diverse historical array of premodern and modern authors. They include Augustine, Maximus the Confessor, and Eriugena, as my premodern representatives, and Friedrich Schleiermacher, William James, and Ralph Waldo Emerson as my modern ones. The term *representative* is deliberately chosen, because it is through them that we glimpse how the focus on "thinking nature" makes for an alternate way of dealing with nature as a theme of religious import. The discussion of *nature* as a religious term has included such questions as the status and meaning of createdness, the impact of God as creator on the world, the role of human nature, and the impact of material reality on thought. Calling attention to an *alternate* way of thinking that I see as being present within the mainstream tradition of Western thought, as I do here, is distinct from seeing this as an *alternative* way in a formal sense. What I am attempting to draw out is not an esoteric, gnostic, or pantheistic tradition underlying the Western religious tradition, or hidden within it, and I categorically reject the attempt to project it as such. I consider all of these representatives to be mainstream thinkers, and however we wish to formulate the parameters of the Western religious tradition, I think that these voices, most notably those of Eriugena and Emerson, my prime interlocutors, should be heard, their weight and position regarded precisely as mainstream, more than they have been. If anything, this book is a plea, and a reason, to do so.

Before I go into the details, dimensions, and historical scope of the exercise of "thinking nature," let me first explain how I came to make this my rubric of choice. Beginning with my interest as a medievalist in Johannes Scottus Eriugena, I have always found medieval thought to radiate something uniquely powerful and fascinating. One of its most distinctive, but also puzzling, characteristics is that it is suffused with religion. As I see it, the religious aspect of medieval thought is an endemic and enduring trait, woven into and inspiring its very structure, which has made me suspicious of philosophical attempts to either dismiss this religiosity as a shell or attribute it to the dated disposition of premodern authors.

But working on the Carolingian Johannes Scottus Eriugena has also put the medievalist that I am in a particular quandary. Not only has he historically been neglected by medievalists, but apart from a small group of specialists, philosophers and theologians have historically overlooked him and continue to do so. Situated between Augustine and Aquinas, in an era that lies fallow between the confessional furrows that have shaped the theological and ecclesial history of the Christian West, Eriugena was already labeled a heretic in medieval times (his *Periphyseon* was condemned in 1225 by Pope Honorius III) for his views on nature, views that would come to be described—when the term emerged in the eighteenth century—as *pantheist*. That is, he was considered to have conflated the identity of God and creation in such a way that creation did not appear to have a lasting existence outside God, while God acquired a this-worldly immanence that was unseemly, a charge that all but guaranteed his removal from any intellectual canon while also preventing any later reinstatement.[1] Yet Eriugena's work encapsulates everything that I find attractive about medieval thought: it is self-enclosed while also allowing for different points of entry. It is accessible insofar as it presupposes little to no prior knowledge and, even while technical at times, eschews estericism. It reveals a command of various pedagogical styles, even as it transgresses the predominance of any specific one. Combining and cementing all these characteristics, Eriugena evokes a powerful overarching vision that is deeply religious throughout. His vision primarily revolves around *nature*, a concept that hovers between the traditional sense of Christian creation and a larger, more evocative sense of reality and the cosmos.

As laid out in the five books of his major work, *On Natures* (*Periphyseon*), which is written as a dialogue between a master and a student, *natura* is the central problem of Eriugena's reception. The reason for this is twofold and exemplifies the vulnerability of medieval thought when we allow it to be ghettoized as medieval, the penultimate step before shelving it definitively. One prong of my twofold reason is that the *Periphyseon* is simply hard to read and track. It puts forth a meandering argument, covering nearly six hundred columns in the nineteenth-century Migne edition.[2] As a single text, the *Periphyseon* is not only excessively long, but its wandering style—the weaving of the various themes and genres that make up its investigation of *natura*—obviates a straightforward reading of the work. In book 1, for instance, Eriugena gives us a discussion of dialectic, the divine names, and the Aristotelian categories, even as he also launches into the meaning of theophany, that is, of creation as divine manifestation. In book 3 Eriugena gives us an intriguing excursion on divine nothingness (*nihil*) before embarking on a literal reading of the creation story of Genesis, and in books 4 and 5 he switches to an allegorical reading of that biblical text. All the while he is engaged in completing the work's intended goal, which is to guide nature on its journey back to God, in whom it originated, as he is keen on complementing the so-called procession with a matching return.[3]

The second prong of my twofold reason for the *Periphyseon*'s neglect is that in contrast to a scholastic work like Aquinas's *Summa theologiae*, the themes and arguments of the *Periphyseon* cannot be divided into discrete compartments that are further subdivided through distinctions without substantively affecting the whole of which they are a part. For Eriugena, the whole of *natura* not only precedes but also always fully permeates each and every one of its parts. As a result, these parts represent and can speak for the whole of *natura* at all times, as befits nature's inherent dynamism, and they cannot be replaced or rearranged without doing lasting damage to them individually, as well as to the whole of which they are constitutive.

Although Eriugena was charged as a pantheist because *natura*—which he divides into things that are and things that are not based on whether or not they can be understood—notably includes God (who transcends understanding and hence resides in nature as nonbeing), I have increasingly wondered why the aspect of divine immanence is necessarily considered so scandalous, given that omnipresence is a traditional attribute of the divine from Augustine to Luther and beyond.[4] An alternative and more plausible hypothesis to the question of why Eriugena's work became theologically troublesome is precisely its indivisibility—in terms of both presentation and content. The indivisibility of Eriugena's work—and by inference the intractability of nature that drives and is glimpsed within it—ran counter to the compartmentalization of Western religious thought that would become standard in scholastic education from the thirteenth century onward and that found a philosophical continuance of sorts in the segmented rationality of Enlightenment thinking.[5]

To bring out the unicity of Eriugena's project then—and here I come to the origin of my chosen rubric—I see the *Periphyseon*, when pared down to its dynamic essentials, as an amazing attempt at "thinking nature." Standing out, in the way it does, as Eriugena's container and schema for "all things," nature acquires from Eriugena a vigor that defies the ordinary limitations of Christian "creation," though he is generally careful not to eliminate the transcendent power of the divine entirely. More important, ingrained in *natura* we find a deep-seated desire to act and move on its own and, especially, to chart its own course, rather than being something more passive that exists simply to execute a divinely dictated script.

It is this quality that has led me to set up the book in the way I have, that is, with the focus on "thinking nature." In my view, thinking is the most crucial and radically free act known to the otherwise closely regimented and hierarchal Middle Ages. With social and gender relations narrowly prescribed, and the church itself a body increasingly known not just for the aesthetics of liturgy but for power and the enforcement of discipline, there is an inherent unconventional robustness to medieval thought that Eriugena's project of "thinking nature" both exemplifies and reinforces.

Natura itself, then, conceived in Eriugenian fashion as the pulsating whole of reality to which God is interior, is where this book finds its deepest inspiration. In trying to approach the intractability of Eriugena's nature, I bring him in conversation with Ralph Waldo Emerson, whom I have come to appreciate as the other master of "thinking nature" in Western religious thought. Emerson, it is good to know, is not just helpful in this conversation because he is located outside the medieval period. In fact, I chose him over his contemporary, Thoreau, precisely because of the near-medieval robustness of his thought, his willingness to mix genres, styles, and even audiences. He is quite able to project a power and vision entirely his own. This has caused significant reception issues, as he has alternately been considered a pamphletist, aphorist, lapsed Unitarian minister, and nineteenth-century motivational speaker rather than, indeed, a thinker, which is how I choose to see him. Though no ties bind Eriugena and Emerson—theological, philosophical, or otherwise—what joins them at the hip is their uncommon enterprise of wanting to "think nature"—that is, engage in thinking in such a way that they hold nature in their minds, always reserving space for nature's ability to dictate thought, all the while recognizing the innate correspondences or links between nature and selfhood. Both authors "think nature" wholesale, that is, apart from any looming threat of instrumentalization—which in Emerson's age of budding sciences was always just around the corner—and with regard for neither prior models of framing nor forensic means of compartmentalization. In fact, both thinkers concur in that they abstain from any desire to frame it at all. Like Eriugena, Emerson does not buy into firm dividing lines separating what is human, what is created, and what is divine. The only principle to which both defer is that of seeing the maieutic self as nature's sparring partner, its interlocutor rather than master or slave, its midwife but at the same time its offspring. Human selves are, after all, created beings or, as Augustine says of himself at the opening of *Confessions,* "a portion of your creation" (*aliqua portio creaturae tuae*).

This commonality of principle, in the robust medieval way that I see instantiated as vibrantly in Emerson as in Eriugena, means that, of their vision of the connection between self and nature, we can also say the reverse: that is, see created beings and nature as if they are human selves. In fact,

Emerson is keen to engage precisely in this kind of wild and daunting hyperbole, stating as he does that "plants are the young of the world, vessels of health and vigor; but they grope ever upward towards consciousness; the trees are imperfect men, and seem to bemoan their imprisonment, rooted in the ground" ("Nature," *CW* 3:105–6).[6] His is truly a whole and undivided universe, then, suffused with religion in spite of his departure from the ministry, where "nature is thoroughly mediate" and "the farm is a mute gospel" (*Nature, CW* 1:25, 26).

So far, so good, then, insofar as we have now established a connection between nature's dynamics and its indivisibility-cum-intractability, perhaps even irreducibility. In Emerson's "Method of Nature," however, he goes further still by describing nature as "the best meaning of the wisest man" (*CW* 1:132). Following Emerson's lead, I see the project of "thinking nature" edge closer here to another kind of reversibility, allowing for the idea of "nature (doing the) thinking" in addition to "(a person who is) thinking nature." What I mean by this is a concept of nature that is fully set free, released from the constraints of human, or even divine, control and that seems to actualize most concretely there where it is able to melt with pure thinking. In such a melting, nature acquires a temporal dimension through achieving or performing what Emerson calls "onwardness," the notion that humans are never completely in control of their own thoughts insofar as nature is always prospective, putting us on our way.[7] It is this deeper assimilation of nature with thought and temporality that the juxtaposition of Eriugena and Emerson allows us to pursue further, whereby thought does not indicate the equivalent of ethereal speculation but instead references the crystallization point that turns the world into a free space where agent and act are no longer distinguishable and where object and subject can trade places.

One of the motivating reasons for me to go this route is that, if we can indeed conceive of nature as thinking, expressed in the notion of a thinking nature, we can then begin to develop an aspirational concept of nature that reclaims agency and empowerment. In such a concept, nature is no longer only passively defined by being subordinated to human or divine conversation—a changing thing, yet devoid of agency—but becomes empowered to take matters (i.e., life) into its own hands, rendering both humans and the divine attentive listeners to it. When nature is no longer just passively conceived, we are invited through the exercise of "thinking nature" to construct a thinking nature. A different and altogether new register of thought opens up, one that Emerson taps into throughout his writings. This is the register of nature as force, as executing a prophetic office, as not only representing "the best meaning of the wisest man" but reaching beyond the passive role of representative. Nature as force projects divine thoughts but does so authoritatively and of its own accord. This concept of nature as force stands quite apart from the charges of pantheism that have been leveled at our two thinkers. It simply reaches beyond the orthodoxy-heresy divide.

To bring all this out, I develop in the first chapter what I call an Eriugenian-Emersonian axis of "thinking nature." To flesh out this axis and preserve the myriad transgressive and hybrid forms that "thinking nature" can assume in their individual works, I have flanked my portrait of the Carolingian and Irish Eriugena with chapters on Maximus the Confessor (580–662) and Augustine (354–430), placed in chronologically reverse order, while I surround the modern and American Emerson with chapters on Friedrich Schleiermacher (1768–1834) and William James (1842–1910). These figures form a diverse array of thinkers. By combining an Eastern (Byzantine) and Western church father with a continental and an American religious thinker, I mean to express my desire for a more capacious canon of Western religious thought. The concrete goals of these flanking chapters are to clarify how Eriugena and Emerson stand out from their age while also showing what these authors share. My hope is that by familiarizing ourselves with Eriugena's and Emerson's concepts of nature within the broader context of a more historically diverse and inclusive set of investigations into "thinking nature," we can take our leave from the conventional notions of creation that have been all but taken for granted. Taking these conventional notions for granted has meant that they have been imposed as uniformly valid and, therefore, have become oppressively representative, which is altogether unnecessary, given the robustness and freedom inherent to both thought and nature. In their place I would like for us to be open to receiving a new, more open-ended and asymptotic, notion of nature, one in which the divine is an integral, rather than a foreign, presence and the self plays a role that is internally constitutive of nature rather than externally reflecting on it.

One unusual but appealing direction in which the new conceptualization-cum-construction of nature can lead us is that of nature as listening (listening because it chooses to, as agent, not because that is its passive function), as compassionate, as assuming a new mantle of stability (as opposed to necessary victimhood), serving as a kind of wailing wall for prayers of despair and disillusionment as we reflect on our present ills and past wrongs. It is in such a conceptualization that I see nature's religious quality play out more fully, a quality that is deep-seated and goes far beyond its coming into being as the delimited product of divine creation. What we see here is nature having an authorized divine mission to serve as a well of compassion, consolation, and insight. In this respect it is not surprising that the source for the personification of Lady Nature in twelfth-century medieval allegories was that of Lady Philosophy, the protagonist of Boethius's *Consolation of Philosophy*.

The idea of nature acting as consoler-in-chief accords surprisingly well with the agentially receptive and deeply intractable character that it exhibits in both Eriugena and Emerson. Both of them see nature as highly malleable—"thoroughly mediate" in Emerson's terms[8]—yet by allowing it to have agency in charting its own course, they also lend nature a quality of being perennially unperturbed. One takeaway from the notion of a compassionate and consoling nature is that it allows for a portrayal of nature that goes beyond both victimhood and inviolability. Instead, the resultant portrayal of nature allows us to project a wide-ranging spectrum of cosmic attitudes on its canvas even as nature maintains its own integrity, an integrity that shows it to be far from indifferent, while forcing us to hold it and ourselves to greater account.

The above lays out the project of "thinking nature" and its various ramifications as it emerges through my juxtaposition of Eriugena and Emerson. But there is another, still deeper, level of "thinking nature," for which I mean this project to serve as a kind of preamble. As I explain in the first chapter, nature is there before we are. In fact, we become who we are only by distinguishing ourselves from nature, gaining our own identity as different and increasingly independent from it. Hence, we might say, in a slight rephrasing of the Augustinian statement above, that before assuming our anthropological selves, we have identity as natural beings, whether or not we choose to put that in Christian creational terms or not. In *The Great Derangement: Climate Change and the Unthinkable*, the Indian writer Amitav Ghosh captures the alignment of human identity with nature when, overseeing his own travels and the journeys of his ancestors, he states: "When I look into my past the river seems to meet my eyes, staring back, as if to ask, Do you recognize me, wherever you are?" Ghosh then reflects on recognition as being the passage from ignorance to knowledge. Complicating the assumed uniformity and forward movement of this transition, he stresses that its most important element lies in its first syllable: *re-*. And he expounds: "The knowledge that results from recognition, then, is not of the same kind as the discovery of something new: it arises rather from a renewed reckoning with a potentiality that lies within oneself."[9]

Transposing Ghosh's thought to this book, I consider my musings on nature quite different from the various environmental projects in science, ethics, and humanities that are so quickly populating the libraries of the contemporary academy. While many of them breathe an atmosphere of crisis, in large part justifiably so, this book wants to take a step back into the historical past, to do the work of the "*re-*," and thereby bring under way the process of the "renewed reckoning with the potentiality that lies within oneself" on the level of Western religious thought. Even aside from any environmental crisis, in the way that nature has been conceptualized over the course of two millennia of Western religious thought, nature finds itself increasingly boxed in between the boundary concepts of *creatio ex nihilo*, on one hand, and pantheism, on the other. This unnecessary and unfruitful delimitation is itself ample reason for it to be rethought. I can here only offer a new starting point with such a reappraisal, but I hope that my attempt to conceive of nature more dynamically and more capaciously will have a ripple effect, affecting our thinking not just about creation but, more broadly, about the role and engagement of religious tradition in the formulation of our concepts.

Obviously, and perhaps inescapably, there is a moral element to this book, not only with regard to the concept of nature animating it but also to the quality of thought that can result from being grounded in such a conceptualization. The more we accord nature a measure of integrity and agency, the more we will find that we need to hold ourselves to account. Schleiermacher and William James feel the need to distinguish between religion and morality and struggle to do so, this struggle for distinction becoming a backdrop against which their views on nature and creation seem increasingly muddled. Part of the difficulty of their discernment process is no doubt attributable to the effect of an encroaching secular mind-set on their thought, which clearly sets them apart from the premodern thinkers treated in Part 1. In this way, though, their struggles underline the uniqueness of the equally modern Emerson. For Emerson, nature's prophetic office is a steadfast theme, and, with a truly natural seamlessness, nature is at once a carrier of religion and truth, as well as of justice. Thus, we encounter Emerson again as a thinker of near-medieval robustness who both represents and transcends his era, not only by looking ahead but also by harking back, pointing to untapped potentialities. He, more than any of his contemporaries, urges us onward to "renewed reckonings."

It is through Emerson, then, that I must read Eriugena. To deal with nature, we need a greater arc of human accountability and a more capacious sense of religious tradition. I find both of these expanded notions most readily in Emerson in a way that, once excavated, we can see more clearly in Eriugena. The aim is to expand our familiarity with this fresh notion of nature so that we might achieve greater intimacy with nature itself. Thus, while it is in Eriugena that the project of "thinking nature" has its origin, it is with Emerson that its narrative will commence.

I

Thinking Nature in Eriugena and Emerson

Thinking Nature

This book revolves around the problem of how to think nature—that is, how to encounter nature in thought, bestowing identity and agency on it rather than objectifying it by means of thought. In the Western religious tradition nature has, on the one hand, been too narrowly identified with Christian "creation," which is placed opposite God, and, on the other, too often regarded as a foreign force or immoral entity rivaling God and hence something to be shunned or kept at bay. As a result, two streams of nature discourse run through the Western religious tradition: the discourse of exegesis and ongoing theologizing of the theme of biblical *creation*, on the one hand, and a more elusive sense of *nature*, which is seen as wily and wild, transgressive and unbridled, and at times considered tinged with pagan (not to mention feminine) overtones, on the other.[1] Insofar as the more elusive *nature* concept is seen as trespassing on sacred ground by threatening to overwhelm and sully the more orthodox, submissive *creation* concept, it has also become labeled as pantheistic. *Nature*, in this sense, is considered neither safely under divine control nor fully penetrable by the human intellect.

Given the otherness of an untamable and uncategorizable nature, what do we stand to gain by engaging this concept as something integrally connected to the power of human thinking? Is it possible to conceive of a free and dynamically operative nature that would not merely permit us to encounter it by means of human thought but would even welcome and cooperate with such an overture? And if such an encounter were possible, how would "thinking nature" affect the nature of our thinking other than by plummeting it into a pantheistic tailspin?

The subversive energy evoked when we turn our attention to *nature*, instead of *creation*, has made me increasingly suspicious of the pantheism label. This knee-jerk reaction suggests an estrangement of the human being from nature that is in every respect the opposite of the genuine encounter with nature's otherness to which this project aspires—that is, as not-God and not-self. Nature's status as not-God and not-self, though connected to both, opens the door for approaching it as a dynamic entity—seemingly all-encompassing yet never fully enclosing or delimiting the divine—to be imaginatively thought through. For this, a reflexive self or subject and a moment of concentrated attention are required. Even Spinoza's radical adage *Deus siue natura* in his *Ethics*,[2] by which the Dutch Jewish philosopher lays out what is arguably the most overtly pantheistic view of nature in the mainstream Western religious canon, does not mask that he is mindful of the perspectival distinction between *natura naturans* and *natura naturata*.[3] In a measure directly relevant to my project, I see Spinoza's perspectivalism giving God an active stake in both the materialist *extensio* and spiritual *cogitatio* rather than overriding or undermining this Cartesian division, which had become and continues to be hegemonic.[4]

For the far less radical cases of a nature closely aligned with the divine that I will analyze, neither the pantheism nor panentheism label fits. The pantheism label is simply too crude to assess nature's complex and subtle alterity. Far from revealing a lack of divine transcendence or inadequate morality, nature's dynamism reveals an openness that is associated with the divine but that does not fully absorb the divine because of its phenomenological uncontrollability. Nor can such a dynamically conceived nature be fully permeated by the divine, which makes the panentheism label equally unsuitable.

What this book argues is that, befuddled at times by a misguided worry about pantheism (or its diluted companion, panentheism), we have neglected the task of articulating what this elusive sense of nature actually stands for, and we continue to do so at our peril. I consider this to be true inside Western religious, as well as secular, spheres and specifically inside Christian circles. There is now a burgeoning interest in developing a new natural theology—brought on jointly by the fading of Karl Barth's emphatic *No!* posited in opposition to the Romantic bond between nature

and religion found in Schleiermacher's liberal Protestantism,[5] and the intellectual confusion caused by the current environmental crisis[6]—which
compels us to take a second look at nature. I see it as dangerously naive to
think that periodic reinforcements of the concept of biblical creation will
do the trick. We live in a deeply unbiblical era, and it seems unlikely that
there will be an imminent return to scripture, either as the former pillar of
Western civilization or as an unfailing oracle for the good.[7] But sacrificing
the integrity of the religious enterprise is no responsible alternative. Such a
step would not only force religious thought to give up its own history and
domain but would also jettison its inherent task to always press onward,
even when faced with unpredictable outcomes, on its way to a more justly
integrated world.

From another angle, the elusive sense of nature conveys that, religiously
speaking, nature is in many ways what comes first. Nature is always there
for us to see, feel, and touch, long before we develop a sense of self through
seeing our reflection in it and, ultimately, becoming alienated from it. It
is as if nature is first a mirror in which we learn to recognize ourselves for
who we are before it hardens into a wall off of which we can bounce ideas.
In yet another familiar image, nature is also a veil through which we sense
the presence of the divine as very close to the touch yet infinitely beyond
reach.[8] Taken together, the three metaphors of mirror, wall, and veil all regard nature, however elusive, as being a meaningful conduit with ties to
both God and the human self. I have been particularly intrigued by the
overarching role of nature as projecting a sense of wholeness—at once expansive and intimate. Exhibiting the traits of both moral compass and eschatological signpost, nature allows us not only to gain a better vision of
God and self but, above all, a better vision of our thought: nature helps us
to see that whatever we think must always be provisional and is always only
one perspective. Hence, while it is there before we are, nature is also what
comes last.[9] Its wholeness, as well as its status as bookending of all being, is
what makes "thinking nature" such an arduous and drawn-out affair.

To be true to nature's integrative wholeness, holding together its many
creases and folds in a seamless garment, I do not want to overemphasize its
distinction from creation, which would only elicit further binaries like Jewish versus Christian, biblical versus pagan, theological versus philosophical.
By "thinking nature" I mean to engage instead in an open-ended intellectual
experiment in which we let nature in its dynamic guise shape our thought

rather than have it fit prefabricated categories, be they religious, theological, or otherwise. Along the way, the attempt to "think nature" will allow us to critique nature's problematic track record in the religious tradition of the West, in which the ingrained habit of thinking *about* nature—that is, treating it primarily as a reified object, unconnected to the human self and mind, as well as to the divine—has too often held sway. Although it is tempting to explain the latter as a result of nature's long-standing religious subservience, which would narrow the problem to an Enlightenment misappraisal of premodern Christian views, I consider the problem of how to articulate and communicate nature's expression, including its self-expression, to be just as prevalent in the modern secular West.[10] In this first chapter I will further clarify what I mean by the divide between "thinking about nature" and "thinking nature" based on the underlying conviction that, religiously speaking, it is imperative that the latter sustain the former. To that end I will engage in a dialogue with "elusive nature" across historical periods. Through this diachronic approach I intend to analyze insights that have accrued over time to develop what I would like to call a sense of nature from the inside out—that is, reading nature on its own terms, which is the opposite of reading it through imposed categories.

Bypassing another binary, this book is neither about nature in its facticity, as if any viable conceptualization of nature must ultimately revert to a responsible ecoscientific course to get things right,[11] nor about human nature in its thrownness,[12] as if only postmodern Heideggerian terminology can bring the tortuous arc of Western naturalist thought to a close, unmasking in the process how it has been polluted by inauthentic Christian values. As I see it, since both scenarios—the scientific-environmental and the human-deconstructive—equate nature's religious self-expression too quickly with an assumed state of victimization, they lead us down a path whereby humanity is inevitably seen as dominating, determining, and domesticating nature. Withstanding the impulse of anthropological guilt, I prefer instead to embrace nature's variegated self-identity as both crude and subtle, transcendental (if not transcendent) and earthly, enchanting and material, yet fully religious throughout.

Given its inherent panoramic quality, for our natural habitat always surrounds us, nature precedes abstraction, but, as the near limitless universe that it is, it seeks at the same time to be recognized, its omnipresence demanding to be channeled. As a result, nature inserts itself indelibly into the human imagination, all the while resisting preconceived models of thought and, in the end—and I see this as a key ingredient of nature's elusiveness—remaining irreducible and intractable. Insofar as I want to do justice to nature's inherent intractability, the project of "thinking nature" must perforce engage the human self. I have come to prefer the term *human self* to alternatives like *human nature, soul,* or *humanity,* a choice that I see as being in line with my attempt to analyze nature from the inside out. Yet despite the apparent similarity between nature and self—does the human self not also elude definition and abstraction, which it likewise precedes?—there remains an antecedence to nature. Rubbing against the natural world is the first step the human self makes toward finding itself *qua* self, their distance from each other a prerequisite if the self wants to make nature its home. Ironically, nature's intractability, coupled with its ambiguous religious status, as we will see, may well help to explain why the history of Western nature, its aspect of *longue durée,* has not yet received the same in-depth analysis as the human self, given that to date there has been no naturalist Freud.

Situating this book's first part as I do in the discourse of the premodern West, we will find nature's character there predominantly conveyed by the Christian term, *creation,* reflecting how God called nature into being through the Divine Word.[13] But by restricting ourselves to the biblical term *creation,* we would not only risk cutting out nature's intractable aspects but also premise its interpretation too quickly on what we assume the authority of the divine word, that is, scripture, to be.[14] By contrast, engaging in "thinking nature" rather than conceptualizing creation allows us to fully honor nature's antecedence to the human self and its intractability, while including the dynamic range of human selfhood allows us to widen the parameters of religious debate beyond the canon of biblical literature. The latter is all the more necessary given that in traditional theologies of creation, the role of humanity is too often reduced to either its laudable status as the image of God (*imago Dei*) or its external, executive, and almost mechanistic office of biblical stewardship.[15]

An even stronger reason not to make *creation* this book's central concept is that it would muffle the religious voice that I consider nature's own. All too often, it seems, the tense creator-creation binary makes scholars fearful of wresting control away from the divine. As an example we may look at theologian Kathryn Tanner's insistence that creation (or human community for that matter) not be construed as competitive with the divine,[16] as she counsels us against inadequately conceptualizing divine transcendence.[17] Insofar as she emphasizes a nonexclusive relation between transcendence and immanence as two divine modes of being, attempting to safeguard both divine and human freedom in the process, her plea for God's noncontrastive transcendence deserves to be heard, even though its chief theological gain does not reach beyond reiterating established tradition. Eager to strip away any and all religious ventriloquism, my attempt to get at nature from the inside prompts me to abandon the guardrails of formal theological structures and let nature speak for itself.

Letting nature speak for itself is also what various twelfth-century allegorists did when they personified nature. One may think here of Bernard Silvestris's *Cosmographia*, in which he projects the mythological emanation of macro- and microcosm from Physis, Urania, and Natura onto his French contemporary landscape, or of Alan of Lille's prosimetric *Plaint of Nature*, in which Lady Nature accuses humanity of sexually abusing her and threatens it with excommunication.[18] Elevated to semidivine status, Alan's Lady Nature echoes Boethius's Lady Philosophy, who likewise exercised a teaching role.[19] The literary success of these poems notwithstanding, as Bernard's poem was read to Pope Eugene III in 1147 to great acclaim, both nature's generative capacity—what Alan calls its production of "like from like"—and its emancipated status as semidivine teacher so worried religious traditionalists that a default theological reaction to curtail nature's free expression and rein in its religious initiative was hard to avoid.[20] As a result, nature in the postmedieval Western religious tradition oscillates between wanton expressiveness, on the one hand, as is evident in the nineteenth-century Romanticism of Hölderlin and Schelling that develops contemporaneously with Schleiermacher's thought, and undue restraint, on the other, as neo-orthodox critics like Barth placed nature under a biblical gag order.[21] Rather than seeing nature's profile in the West as tainted by traces of paganism—which was Barth's ostensible worry in his debate about natural theology with Brunner—I see it marked by a centrifugal channeling of its religious claims owing to an insufficient regard for, if not fear of, its intractability. Seen as intractable, nature offers neither compass nor memory, but by impelling us to reflection, it affords the self the valuable opportunity to become more firmly anchored in reality.

In premodern Christian accounts, nature's range is often found delimited by the doctrine of *creatio ex nihilo*. Although not expressly designed to curtail or suppress nature's religious voice—as I will clarify in the chapter on

Augustine—*creatio ex nihilo* appears to have been increasingly used for that purpose, as dependency on the divine morphed from existential premise to reified attribute. That *creatio ex nihilo* conveys forcible restraint is clearest in the early Christian and Reformation periods and is a recurring concern of the Barthian tradition. In Barth, *creatio ex nihilo* is not merely an antidote for pagan influence but oddly became the chief argument launched *against*—rather than *in support of*—natural theology altogether, as Barth considered natural theology to lack sufficient deference to divine transcendence. This reveals Tanner's insistence that creation not compete with the divine as at root a Barthian one.[22]

Fear of pantheism appears to have been the culprit of such theological rigidity, particularly from the Enlightenment era onward. Given that I see nature's identity confronting us more with elusiveness and intractability than with heresy, I see the pantheism charge as a random, if not altogether empty, accusation that aims more at clarifying the role of the divine than of nature.[23] Comparable to how theodicy, as a positive term of orthodox Christianity, became used to promote divine omnipotence,[24] the pantheism charge seems to have been largely leveled for the purpose of reining in a breakaway, quasi-independent nature by enforcing the sovereignty of God.[25] To the extent that the effect of such a charge was to silence nature's religious voice, I want to call it out for the original sin against nature that it is: a misguided and mendacious attempt to disempower nature—traditionally cast in a feminine role—by defertilizing it.

Invoking the pantheism charge here, with panentheism as an unhelpful surrogate, makes plain that tracking nature's residual otherness in the religious tradition of the West is indeed a difficult and complex task. Considering nature beholden neither to the ecoscientific objecthood of modern science nor to the philosophical thrownness of postmodern deconstructive theory, and unsatisfied by its equation with biblical creation, I will tease out its religious role by engaging it in dialogue across historical periods and, more important, across the chasm that severs the premodern from the modern religious tradition. To that end I present a series of historico-religious case studies to serve as signposts for what the sustained practice of "thinking nature" might entail, including how it might affect the human self. In the process we will design a corresponding vocabulary for thinking—and talking—nature, as a condition for embarking on a deeper discussion of its innermost movements.

Eriugena and Emerson

As my reference to twelfth-century allegory indicates, attempts to "think nature" are not without precedent in the religious tradition of the West. There have always been thinkers ready to assign nature a place and lend it a voice, even at the risk of religious criticism or marginalization. Given such attacks, it is important to realize that the major hurdle hampering our understanding of twelfth-century nature poets is not their perceived heresy but their stated ambiguity.[26] Inspired to express the full range of nature's endemic powers, but aware of the risks involved in broadening the boundaries of existing tradition by including non-Christian sources, Bernard and Alan made subtle use of the trope of *integumentum*, the idea that pagan myth is a wrapping around an ultimately Christian truth.[27] Doing so allowed them to inject polyvalence into nature's articulation of its role. While this made them sophisticated poets, it caused the religious quality of nature's emancipated state to fall by the wayside. Thus, when, after the founding of the first universities, scholastic humanism—to adopt the term coined by R. W. Southern—became the new paradigm of higher education in the West,[28] nature largely stopped speaking outside of the realm of poetry. From the thirteenth century onward—with the exception of various Renaissance Platonists considered too idiosyncratic to be included in any canon—the analysis of nature (read: "thinking about nature") and the facilitation and channeling of its self-expression (read: "thinking nature") would proceed largely as separate undertakings.[29]

My dialogical engagement of nature strives to return the aspect of nature's self-expression to the analysis of nature, for which my selected case studies can serve as a budding canon. Two thinkers whose work I consider particularly pertinent to the theme of nature's self-expression serve as my primary interlocutors in this endeavor—namely, the early medieval Irish intellectual Johannes Scottus Eriugena (810–77) and the nineteenth-century American thinker Ralph Waldo Emerson (1803–82). While both thinkers maintain nature's pride of place and attendant religious role, they also share the idea that, if the enterprise of "thinking nature" is to be respectful of nature's self-identity, it must elevate nature above its generative function and bring out what most deeply animates it or, better, what makes it tick. In other words it is the insight into nature-in-action—that which twelfth-century authors indicated with the term *natura operans* (operative nature)[30]—that qualifies

Eriugena and Emerson as suitable conversation partners who can help us not only launch the dialogue with nature but sustain it as well.

Since I see the attempt to "think nature" largely as a heuristic and hermeneutical quest—a presupposition that I believe my two chosen interlocutors share—it will require considerable navigating skills to track nature's movements with precision. As we have seen, nature always finds itself in flux between making, self-making, and being made (or created), inhabiting various roles but never wholly settled in any of them. My hope is that by analyzing Eriugena's and Emerson's attempts at "thinking nature" in more depth, even if we need to straddle the medieval and the modern era to do so, we are able to plant useful buoys that will help us point a meaningful way forward toward a new and reinvigorated natural theology. While I will bring in more authors to amplify our fledgling natural canon, it is around the Eriugena-Emerson axis, where nature is positioned among making, self-making, and being made, that each of these thinkers finds a place.

Before discussing the individual cases of Eriugena and Emerson, however, representing as it were the two poles of my ellipse, let me explain more fully why I chose them to be my guiding thinkers. Although the turn to a medieval thinker in a study of nature need not in itself be surprising, my preference for the Carolingian Eriugena over, say, one of the twelfth-century allegorists, or the emblematic Franciscan Bonaventure, may be.[31] My reason is that the consistent attempt to "think nature" pushes Eriugena's thought outside conventional medieval confines, allowing him to transcend the medieval era even as he continues to inhabit it.[32] In defining Eriugena's eccentricity, I have found him to resonate most productively not with any of the above medieval thinkers but with the nineteenth-century American author Ralph Waldo Emerson. In the absence of direct source influence, as there is no evidence that Emerson ever read Eriugena, their remarkable convergence rests on what I regard as a similar eccentricity in Emerson, whose break with the church and university of his day led him to unleash the dormant potential of American thought. It is by capitalizing on the eccentric profiles of both these thinkers, representing as well as transcending the cultural framework of their respective eras, that we may gain a thicker notion of what "thinking nature" entails.[33]

Pushing yet further, I contend that Emerson offers us a more straightforward entry point into the ambiance of Eriugena's nature than reading Eriugena himself does, given that his long and winding *Periphyseon* is exceedingly difficult, if not obscure. The application of an Emersonian filter can help us unpack Eriugena's thought by stripping it of the convoluted Neoplatonic trappings that were "Christianity's content" during much of the premodern era and newly integrate its composite parts.[34] A more difficult question, but one I nevertheless want to pose, is whether the reading of Eriugena can also benefit the analysis of Emerson and, if so, how. This is especially relevant since the latter has in recent decades been increasingly transformed from an American master of literary aphorisms into a capacious thinker in his own right, one particularly sought out to overcome the philosophical stalemate between analytical and continental thought. For Stanley Cavell, who perennially wrestled with skepticism and reality's elusiveness to thought, and who also played a key role in this transformation of how we think about Emerson,[35] Emerson's attraction lies in his casual acknowledgment of existence. For Cavell, Emerson's acknowledgment of existence offers a more effective answer to Descartes's question of the "I" of the subject than Kant's retreat into the noumenal world, insofar as it compounds any easy access to (a schematized) reality.[36] While much of what Cavell derives from Emerson pivots on the human self, given that Cavell aims to return contemporary philosophy to its originary task of philosophizing or, rather, of prephilosophical conversation,[37] he rarely thematizes Emersonian nature aside from its acknowledgment by the moral self.[38] Projecting the Emersonian landscape onto an Eriugenian canvas, as I propose to do here, may be helpful precisely in that it not only endows the centuries-old paint of Eriugenian nature with newly discernible hues but also provides useful backlighting to make Emerson's portrait shine more fully. Thus, aspects of the moral perfectionism debate can be transposed to the sphere of a nature that is self-contained yet open-ended, transcendental yet no less material, a conception of nature that has generally been associated more with Thoreau and others.[39]

By regarding Emerson's proclamation of America's intellectual emancipation thus as a palimpsest rather than a creed, we will be in a better position to detect vital and long-lasting ideas on nature beneath its modern surface, even as he makes them newly visible. After all, the rehabilitation of Emerson as a major force in contemporary thought, set in motion not only by Cavell but also by the work of Lawrence Buell and Branka Arsić, does not assume Emerson's natural expressiveness to come out of American thin air.[40] Even if Emerson rejects the force of tradition outright, it is not thereby an un-Emersonian move to point out that his thought on nature is tracking a deeper religious groove. It is my wager, therefore, that Emerson's nature gains both focus and intensity when read against the background of Eriugenian nature.

Let us at this point commence our cross-pollinating reading of Emerson and Eriugena in earnest. I will do so in chronologically reverse order. Making the more accessible thinker elucidate the more obscure one allows us to move progressively backward, so to speak,[41] as our reflections on Emerson retroactively enrich the musings of Eriugena.

Emerson on Thinking Nature

As a religiously informed author from the recent past, Ralph Waldo Emerson speaks about nature in a way that seems to forgo crisis. Regarding the moral self as grounded in nature, he seems ideally situated to help us "think nature" without surrendering its fate either to apocalyptic dissolution or unbridled control by scientific, biblical, or theological-philosophical forces. Rather than evaluating Emerson in the more traditional approach as being either a transcendentalist or a pragmatist, I focus on carving out nature's religious role. To that end I have organized my analysis of Emerson's "thinking nature" around two insights from his "Divinity School Address," amplifying them by elements taken from his book *Nature* and essay "Circles."[42] The first insight to which I want to draw attention is Emerson's regard for "preaching," which he defines as the frank speech of man to men. Although he calls preaching the second advantage of Christianity, I treat it here first because it helps us to set up rules for a frank conversation about nature inside the Christian tradition. I treat the first advantage of Christianity—the Sabbath or jubilee of the whole world—after preaching because it positions nature in relation to whatever the perceived apex of history may be; hence, it clearly reaches beyond the Christian tradition. Puncturing any naiveté about nature as a carefree *locus amoenus* or protological paradise, Emerson's Sabbath answers the demands of an eschatology that borders on the secular, since any overt allegiance to Christian tropes has been dropped, and that is entirely devoid of any overtones of apocalyptic crisis.

Preaching

Toward the end of his forceful and highly critical "Divinity School Address," from July 15, 1838, which resulted in his banishment from Harvard for the next thirty years, Emerson settles on preaching as Christianity's, and his, discourse of choice. This is surprising, for the "Address" unfolds, in many ways, as a tirade against preaching. Emerson tells us how, seeing through the window beyond the minister, "the snowstorm was real, the preacher merely spectral" (*CW* 1:85), after which he concludes that this preacher had obviously not learned the "capital secret of his profession, namely to convert life into truth" (*CW* 1:86). Yet Emerson is known to echo and replicate the New England jeremiads that appear to have influenced him, even as he ultimately deems the world of Puritan homilies lifeless and leaves them behind.[43] As much as Emerson's performative act of religious self-immolation—for speech and event appear to collapse in the "Address"—might suggest he had given up on the Christian religion, his attentiveness to preaching is intriguing in that it points to the contrary. As I see it, "preaching" for Emerson is the particular kind of essay—taken here in the sense of an intellectual exercise—that is best suited for what he calls the conversion of "life into truth."

In the "Address" Emerson presents preaching as an important remedy for the flaws of historical Christianity. Insofar as preaching has found itself on a path pointing away from, rather than toward, truth, he reviles historical Christianity. The preference for preaching is indeed counterintuitive at first, since his most scathing indictment is reserved for the preaching he has encountered and the formalism it entails: "it is still true, that tradition characterizes the preaching of this country; that it comes out of the memory, and not out of the soul; that it aims at what is usual, and not at what is necessary and eternal" (*CW* 1:87). But just when one expects Emerson to continue in this antitraditionalist vein, his diatribe gaining momentum, he inverts his position: "and what greater calamity can fall upon a nation than the loss of worship?" (*CW* 1:88–89). Since, for the Unitarian minister that Emerson once was, worship would probably have revolved around preaching, this comment amounts to an oblique acknowledgment of tradition.

Shunning confessional language, Emerson makes clear that he may want to heed tradition but only insofar as he deems it capable of speaking to us "without dogmas." Putting his finger on the pulse of Christianity's historical stagnation, he asserts that "they mark the height to which the waters once rose" (*CW* 1:87).[44] While this may seem to resonate with Friedrich Schleiermacher's view of creeds as concretized historical instances of religious affections,[45] Emerson's take on tradition could not be more different. Both the Protestant Schleiermacher and the Roman Catholic Cardinal Newman saw creeds and dogmas lodged safely within the history of Christianity as beacons measuring progress and stabilizing ecclesial preservation.[46] Not so with Emerson, who only heaps scorn and pity on the preacher transmitting that tradition: "Alas for the unhappy man that is called to stand in the pulpit, and *not* give bread of life" (*CW* 1:87). While dogmas, as if marking the rising waters on an imaginary quay, may be heroic feats etched in the memory of those who survived the flood, dogmatic tradition can never amount to more than a stopgap of sorts. Belying any august solidity, any enduring deposit of faith, Emerson considers its legacy as inconstant as any high tide, which nature—a term not invoked by chance here—will inevitably force to come down. Lacking a foothold not only in tradition but also in the soul, dogmas are by definition ethereal for Emerson; they neither represent nor offer lasting sustenance.

Given that Emerson radically cuts off the high road of ecclesial tradition, the only way left for him to transmit the desired conversion of life into truth is by traveling down the winding road of nature, that is, the road of true preaching: "Where now sounds the persuasion, that by its very melody imparadises my heart, and so affirms its own origin in heaven?" (*CW* 1:85). The rich discourse and vocabulary of nature, accordion-like in its ability to absorb historical linearity yet, on demand, spinning out individual moments, allows him to go both deep inside ("imparadise the heart") and out and up on high ("origin in heaven"). Placing at the preacher's disposal a full repertoire of emotions and colors, nature rather than historical tradition is the ally that can set the proper tone ("melody") for the conversation between God and humanity. Preaching that stands in opposition to nature—compare the spectral preacher whose nearly defunct form stands in marked contrast to the fresh snowstorm outside—only prevents congregations from being nourished by the bread of life.

In light of Emerson's Unitarian distance to liturgy and sacrament, the mention here of the bread of life should not be read as a eucharistic reference; it more likely indicates what we might call the discourse of life, the paraenetic persuasion encoded in good preaching. With a playful homonymy that is hard to imagine having escaped its author,[47] the "Address" evokes a tantalizing contrast between the bread of life as withheld by traditional preachers and "the breath of new life" (mentioned toward the conclusion of the "Address" [*CW* 1:92]) that only true preaching can accord. Galvanizing this connection between bread and breath is the sense that nature, rather than history, or even Nietzschean genealogy, is what infuses Emerson's predilection for "that frank speech of man to men" that expresses and reflects true communication from and with God.

Genealogy was Nietzsche's prime tool to unmask the stolid yet skewed formalism of historical development,[48] allowing him to substitute knots and junctures for the continuous narratives of Schleiermacher and Newman.[49] For Emerson, only nature can bring us back to our original calling, showing tradition to be the wholesale fraud that it has become. Indeed, only nature has the rare and counterintuitive ability to make us turn inward on ourselves. For, as Emerson forcefully puts it, it is not the church but the soul that must be preached.[50] If historical Christianity is to continue—and despite Emerson's personal defection, there is no reason to think he deems it should or could not—what it ought to preach is the rejuvenation of the soul through the breath of new life, which is none other than nature as locus for the encounter of the human and the divine.

Nowhere does Emerson express the urgent rejuvenation of the soul that nature represents more powerfully than at the opening of his book *Nature*: "Our age is retrospective. It builds the sepulchers of the fathers. It writes biographies, histories, and criticism. The foregoing generations beheld God and nature face to face; we, through their eyes. Why should we not also enjoy an original relation to the universe? Why should not we have a poetry and philosophy of insight and not of tradition, and a religion by revelation to us and not the history of theirs?" (*CW* 1:7). Even as *Nature* proceeds by giving us an outright rejection of history, underneath his tirade in the "Divinity School Address," Emerson still wants to acknowledge the continuity of the Christian tradition and, to the extent that it contains seeds of what he elsewhere calls "onwardness," its enduring ability to inspire by communicating the breath of new life. Yearning to reboot what he can only dismiss as calcified religion, Emerson seizes on good preaching as the best way to actualize the premise of Christianity's original revelation. "It is the office of a true teacher to show us that God is, not was" (*CW* 1:89), our reception of whom should hence lead to the transformative imitation of Christ. What he thereby wants to inculcate, however, is neither reverence for the God-Man of atonement, which would be yet another dogma, nor the adulation of other religious heroes or saints but rather a new self-awareness rooted in nature. Through it every human can come to affirm and embrace that "I also am a man" (*CW* 1:90).

It is perhaps because the Christian tradition of preaching is more commonly associated with history as unfolding revelation than it is with nature that Emerson's religious voice in the nineteenth century did not resonate. This failed reception is further explained by the fact that he finds himself at considerable distance from German Romanticism, which was the viable alternative to calcified dogmatic Christianity of the age that did enjoy a warm reception and so marked Schleiermacher and others. Thus, the questions arise of, first, what he stood to gain by carrying on about Christian preaching in naturalist terms, as appears to be the case in the "Address," and, second, why that should impel us to turn to him in our endeavor to "think nature."

Harking back to an earlier point, I consider Emerson's preference for nature over history as the more powerful sparring partner to be ultimately rooted in his conviction that nature can accommodate shifts in perspective in a way that the linearity of history or tradition cannot. This is not to deny Emerson's faith in progress, which he considers one of the great virtues of "onwardness," but nature clearly allows him to open up many more cognitive dimensions of the soul or self, and it can do so all at once. It is as if Emersonian nature outfits us with a wide-angle lens to take in reality wholesale.

The opening scene of the "Address" (*CW* 1:76) presents a particularly interesting instance of nature's kaleidoscopic powers: "In this refulgent summer it has been a luxury to draw the breath of life. The grass grows, the buds burst, the meadow is spotted with fire and gold in the tint of flowers. The air is full of birds, and sweet with the breath of the pine, the balm-of-Gilead, and the new hay. Night brings no gloom to the heart with its welcome shade. Through the transparent darkness the stars pour their almost spiritual rays. *Man under them seems a young child, and his huge globe a toy*" (my italics). Far from putting nature under an impassive microscope, Emerson depicts nature as alive and vibrant. Nature engages both speaker and audience while enveloping them all with its singular multifaceted beauty, even as it draws them along from day to night. Following such well-rounded perfection, which seemingly prepares us for the still ascendancy of controlled abstraction, whether religious or scientific, Emerson's discourse betrays a sudden shift when humanity is reduced to a young child, and nature—transformed into a huge globe—made into the toy with which the child plays. Just as a small child lacks the dexterity to kick or throw a ball with precision, so humanity finds itself at a loss before nature's larger-than-life powers. Yet in the spirit of "onwardness," the child cannot but play, even if it lacks an adequate grip. After the immediately preceding image of nature suggesting order and organization, suddenly we find ourselves as readers caught helpless.

Picking up on shifts like this in his *Emerson's Transcendental Etudes*, Stanley Cavell has called our attention to the unsettling connection between what happens casually and what creates a casualty in Emerson.[51] By that he means that random, everyday occurrences can disclose reality's abyss in ways for which life has left us wholly unprepared but that define for us what life is. Whether in Cavell's focus on individual experiences marked by dramatic incisions or in my own broader focus on Emerson's outstretched panoramic landscape rattled by unexpected psychological shifts, I want to highlight as a constant Emerson's interest in ever-altering our visual perspective. Compare how in the previous image he downsized humans with one stroke of his pen from observers or active participants to playful children! Here there is no Emersonian casualty as Cavell would define it. There is only good fortune or, rather, just fortune whose nature is such that one is never prepared for its impact. After all, the child's innocence leaves it guileless, as unprepared as the audience of the "Address" must have been at the time and not unlike how unsuspecting readers of Emerson may find themselves today. Whether or not some of his rhetorical or homiletic play is done for effect, as he builds up his work, case after case, image after image, instance by instance, Emerson's radical perspectival moves never fail to surprise, and their collective effect wreaks kaleidoscopic havoc on the reader. What's more, this havoc is precisely the point, as it lends a permanent unsettledness to Emerson's texts. Yet throughout the dizzying *tour de force* on which he takes us, he displays a remarkably consistent preference—namely, an irrepressible predilection for the small and the ordinary within nature.[52]

If we bring all this to bear on the Christian tradition, Emerson's unexpected clinging to the small and the ordinary is perhaps best explained by the fact that, through the figure of Jesus Christ, Christianity is in fact quite well set up to foster lowliness. Yet, within the context of a calcified tradition, this original impulse can only be brought out through radical ruptures and inversions. Having earlier kindled the self-awareness by which humans are spurred on to follow Christ and, in this following, each person realizing "that I also am a man," Emerson next says of Jesus that "alone in all history, he estimated the greatness of man," and could yet say, "I am divine" (*CW* 1:81).[53] Rather than suggesting a paradox of human and divine here, or worse, Tanner's specter of shared agency, his point seems to foreground Jesus's humanity, his smallness, but underscores it by using the aspect of divinity as an outsize confirmation. Rather than elevating Jesus's humanity, as if abrogating it through divinity, Emerson uses divinity to affirm and ground Jesus as more authentically human, a paradigm for what Emerson calls the soul.

Trapped in the stagnant dogma of Jesus's divinity, the Christian churches are, according to Emerson, "not built on his principles, but on his tropes" (*CW* 1:81). As a result, they inevitably, and fatefully, distance themselves from nature, in which the essential humanity that underlies Jesus's divinity is ultimately rooted. "But the very word Miracle, as pronounced by Christian churches, gives a false impression; it is Monster. It is not one with the blowing clover and the falling rain" (*CW* 1:81).[54] With Emerson shifting the miraculous back to the ordinary here, the divine to the natural and human, we are invited to order these radical shifts of perspective into a sequence of moments, what Emerson in "Experience" describes as a succession of moods.[55] Refusing to see them as breaches or gaps, Emerson aims to bridge these moods or inversions by reinserting the self into nature. The point of such a reinsertion is not to effect a pantheistic or even mystical union but rather the opposite, that is, to allow the self to aspire to better knowledge and insight through a fuller and more transparent grasp of nature as that which oversees all moments and moods. For Emerson, temporality rather than divinity is what drives us to wrestle with these moods. This is not an easy process, as they can be contradictory, or as Emerson puts it in "Circles," "Our moods do not believe in each other. . . . I am God in nature; I am a weed by the wall" (*CW* 2:182). One must deploy and activate the accordion-like quality of nature—which inflates and deflates the self at will—for these moods to become truly connected.

Humans may exaggerate their Christlike status and see themselves as God in nature, or, in another Emersonian expression, one may become a transparent eyeball (*Nature*, *CW* 1:10), serving as a kind of real-life microscope. But nature as the divine's favorite accoutrement will always have a way of cutting us down to size, forcing us to recognize that what we really are is a weed by the wall, a child with a toy.[56] Therefore, nature, for

Emerson, is the closest that we can come to capturing the lived experience of reality as fraught and imperfect yet still whole, integrally filled with hope and potential.

The Sabbath or Jubilee of the Whole World

A succession of moods—never their union or resolution—is all that Emerson gives us, leaving us to contend with the task of ordering them to form the kind of circular progression that characterizes "onwardness." The natural world, he tells us in "Circles," "may be conceived of as a system of concentric circles." Things that seem to be there for their own sake "are means and methods only—are words of God, and as fugitive as other words" (*CW* 2:186). That Emerson calls things "words of God" affirms their medial quality, opening up a link beyond preaching with scripture that will need to be further explored. What it also reveals, as a further thread connecting individual circles as "uni-verses," so to speak, for there is a succession of them, is that, at heart, nature is dialogue or conversation. Nature furnishes us with the stuff of which the conversation between God and self is made, and only by engaging nature can the conversation truly get under way. It is at this very point that "intuition," as Emerson calls it in the "Address," commences, the different cycles of onwardness now having been left behind, and the first advantage of Christianity, "the Sabbath, the jubilee of the whole world" (*CW* 1:92), becomes a reality. What Emerson exactly means by the reference to the Sabbath here, the jubilee of the whole world, is not easy to gauge, precisely because it is unclear whether there is an eschatological dimension at work and, if so, what that would imply.[57]

Inasmuch as religious traditions—and again, I mean to widen the circle to include non-Christian religions—involve dissemination and transmission of religious knowledge and ways of knowing, it is important to recognize that Emerson makes a fundamental distinction between what he calls intuition and so-called tuition or instruction.[58] Only intuition is original, having the function of opening up what is our original relation to the universe. To the extent that intuition is original, however, it is also lonely ("Let me admonish you, first of all, to go alone," *CW* 1:90), for tradition *qua* instruction will always be unable to "convert life into truth." This conversion can only occur once the self has refashioned itself by means of mirroring itself in nature. For "the Sabbath, the jubilee of the whole world," to arrive, for our hearts to be truly imparadised, we must surpass the iterations of individual religious traditions (*qua* instruction) and gain access to the primary intuition that underlies them.

For Emerson, taking this pedagogy seriously means that we attach great weight to our adherence to not only our intuition in relation to nature but also nature's intuition, of which right religion is a faithful tuition. The exercise of "thinking nature," in the sense of following nature's intuition, will guide us up a path that is circular in the peculiar Emersonian sense: neither cyclical nor gyrating but generating ever greater and more momentous generalizations. Every time one moves out of a circle, one enters another, as another "uni-verse" literally opens up. As Emerson puts it in the opening poem of "Circles" (*CW* 2:177):

Nature centres into balls,
And her proud ephemerals,
Fast to surface and outside,
Scan the profile of the sphere;
Knew they what that signified,
A new genesis were here.

The poem indeed lays out the Emersonian path of "thinking nature," in which the individual's following of Jesus Christ ("I also am a man") or any other religion will eventually lead to the opening up of ever more general circles or "uni-verses." The question here, however, is whether such a path of ever greater generalization can indeed bring us "the Sabbath, the jubilee of the whole world."

In response to that question I will first make two points relevant to the religious and theological status of nature to be discussed in later chapters. By way of preface I want to state that we do better to see the various stances in the fraught contemporary religious debate about nature—biblical stewardship, environmental crisis, and so forth—in Emersonian terms. That is, these stances can be seen not as contrary positions but as orbiting circles, each unfolding according to, but also held back by, its own laws. Emerson's approach would not be to try to perfect or resolve one or the other of the existing circular models: in theological terms, by revisiting the Christian notion of biblical stewardship or reviving the Franciscan engagement with creation;[59] in environmental terms, by coming up with responsible scientific solutions to avert ecological catastrophe.[60] Emerson's intuition is that we need, rather, to roll out a new circle, to engage in a reshaping of the current human relationship with nature that prevents such bifurcations.

Thus, having invoked the model of the Emersonian circles as moving toward ever greater generalization, we can now couch with a bit more clarity the canon-building work that this book aims to begin. I intend for the following chapters to unfold like a slow-moving circle: one that bends the Eriugenian-Emersonian axis developed here into a more holistic canon or, better, a roundtable and symposium. To do so, I gather Maximus and Augustine to the table to complement my reading of Eriugena with chapters that focus, respectively, on nature's ties to incarnation (Maximus) and nature's ties to scripture (Augustine), followed by a postscript on nature as conversation in Eriugena. After a short transition from the premodern to the modern, Friedrich Schleiermacher and William James join the roundtable, as I extrapolate their views on nature, religion, and the self. In my conclusion, I integrate the foregoing chapters in relation to the central axis of our burgeoning symposium, as I draw a final, more general circle to bring out the meaning and reach of elusive nature.

We will now return to my two points on the Sabbath. My first point is that Emerson's "thinking nature," as a way of conceiving nature through encounter in thought, allows us, internally and integrally, to deepen our thinking of nature *to include the self.* Notwithstanding his definition of nature as everything "not-me" (*Nature*, *CW* 1:8), there are too many instances in his essays in which nature is conceived as an extension of self, or in which the self is shown to be so enveloped by nature, that there is no place or moment where nature does not also draw in or mark the self—or as Emerson calls it in the "Address," the soul.[61] As he puts it in "Circles," there is one law of eternal procession that not only governs and animates nature but also arranges all that we call the virtues (*CW* 2:186). There, the highest prudence will meet the lowest prudence in the "transcendentalism of common life" (*CW* 2:187). This intriguing "transcendentalism of common life," importantly, makes humans aware that they are not only part of a larger whole but, in fact, constitute the lynchpin of the various expanding circles that constitute nature. Of these expanding circles, Emerson writes, "the first is the eye, the second the horizon which it forms," and the widest is the "nature of God as a circle whose centre was everywhere and its circumference nowhere" (*CW* 2:179). We should note here as an aside that the latter image Emerson mistakenly attributes to Augustine; however, it derives from *The Book of the Twenty-Four Philosophers* and is adopted in a sermon by Alan of Lille on the *sphaera intelligibilis* (intelligible sphere).[62] To quote Emerson directly, from the end of *Nature*: "The problem of restoring to the world original and eternal beauty, is solved by the redemption of the soul. The ruin or the blank, that we see when we look at nature, is in our own eye" (*CW* 1:43).

In Emersonian scholarship there has been a tendency to take the convergence of nature and self so far that there appears to be no problem with nature at all, only with the self. For example, I see Cavell's subtle enduring focus on moral perfectionism, which emerges from his predominantly anthropological reading, as potentially overlooking the inherent complexity of nature.[63] To the extent that the idea of an eschatological Sabbath is meaningful to Emerson (that is, over and above the ordinary weekly day of worship), it must play out against the background of a full integration of nature and self rather than a collapse of nature into selfhood. Emerson writes, "The axis of vision is not coincident with the axis of things, and they appear not transparent but opake. The reason why the world lacks unity, and lies broken and in heaps, is, because man is disunited with himself" (*CW* 1:43). This includes, by extension, the integration of nature and culture or nature and society. That is, in eschatological terms, Emerson here seems to envisage a collapsing of time and place through forging a kind of coinherence of the Sabbath (time) and paradise (place), leading to an enlarged view of the world in which the self is more fully at home with it and with itself.

My second point is twofold, as I try to elucidate the most ingeniously relevant aspect of Emerson's nature for the project of "thinking nature." On the one hand, Emerson has indeed a big role left for the anthropological, promising nothing short of the world to the common man. As he states at the end of *Nature*:

Every spirit builds itself a house; and beyond its house, a world; and beyond its world, a heaven. Know then, that the world exists for you. For you is the phenomenon perfect. What we are, that only can we see. All that Adam had, all that Caesar could, you have and can do. Adam called his house, heaven and earth; Caesar called his house, Rome; you perhaps call yours, a cobbler's trade; a hundred acres of ploughed land; or a scholar's garret. Yet line for line and point for point, your dominion is as great as theirs, though without fine names. Build, therefore, your own world. (*CW* 1:44–45)

To some extent Emerson sees nature here, in the words of phenomenologist Jean-Luc Marion, as a saturated phenomenon,[64] a surplus of intuition over intention, something that indeed includes and brings forth culture, spirit, and soul. But nature also always retains its otherness, to the divine no less than to the human. It should be noted that while nature as saturated phenomenon obviously betrays divine-like qualities, humans betray those same qualities, as Emerson's Christology makes clear, so this issue is not what is really at stake. In the end there appears to be a remarkable transience to divinity in Emerson or, at least, to its appearance in the world. This transience, however, cannot be said of Emerson's nature, which plays a key role in the Sabbath or jubilee precisely because it is what endures.

In sum, nature has the ability to completely absorb the religious and prophetic task for Emerson, as if an office, and as such it is ready and fit to be tasked with teaching us how to live rightly. Nature is tuition and instruction, even prophecy and proclamation. As the only true other to humanity, as well as to the divine, it finds itself in the unique position that it can pronounce judgment, not just on them but also on their mutual relation and the quality of that relation. And yet, as if ecologically aware of this momentous responsibility, Emerson's nature always retains a certain reticence about itself, one that will only truly be broken at the dawn of the Sabbath or jubilee. This reticence is not to be passively accepted, and I do not think Emerson would expect us to do so, but it rather cries out to us, at every instant of every day, to be punctured. Indeed, breaking through nature's reticence is humanity's only appropriate response to the gift of nature's tuitions. As Emerson says in *Nature*: "All things with which we deal, preach to us. What is a farm but a mute gospel?" (*CW* 1:26). Such relentless puncturing lends a voice to nature's mute gospel, and by doing so boosts and supports its mission to be thoroughly mediate.

Eriugena on Nature and the Self

Having gotten a glimpse of the projected images onto the palimpsest, we now turn to the old canvas underneath and what of its colors and contours remain and show through. Switching from Emerson to Eriugena, we enter a radically different milieu and a much earlier era. Eriugena's major dialogue, the lengthy *Periphyseon*, Greek for *On Natures*,[65] dates to the 860s, the so-called Carolingian renaissance. In this unusually powerful, but also hybrid, text he introduces an idea of nature that is as radical as it is original. Owing to the *Periphyseon's* posthumous condemnation in 1225 for, ironically, pantheism,[66] it never had much recorded purchase in Western thought. Marginalized inside as well as outside the Christian tradition, since Eriugena also never received the belated hero's welcome that befell a perceived heretic like Abelard, the *Periphyseon* was excised from the established theological-philosophical canon of the West, and its ideas never took root there.

Yet the work's deliberate and self-conscious foregrounding of nature sets it apart from earlier attempts like Origen's *On First Principles*. Marking an important juncture in the religious history of the West, it presents us with one of the earliest available and most consistently radical instances of what we might call *natura operans*, a portrait of nature in action.[67] Eschewing conventional creation terminology, Eriugena aims to integrate the arc of *natura* fully with the explorations of the human self, steering clear of both human crisis mode and the heteronomy of divine control.

To analyze the *Periphyseon's* vision of the interrelatedness of nature and self, let us look at how its dialogue of Master and Student gets under way:

> Master. Often as I ponder and investigate (*saepe mihi cogitanti*), to the best of my ability, with ever greater care the fact that the first and foremost division of all things that can either be perceived by the mind or transcend its grasp is into things that are and things that are not, a general name for all these things suggests itself, which is *physis* in Greek or *natura* in Latin. Or do you have another opinion?
>
> Student. No. I definitely agree. For when entering upon the path of reasoning, I also find that this is so.
>
> M. Is nature then the general name, as we have said, of all the things that are and that are not?
>
> S. That is true. For nothing at all can occur in our thoughts that could fall outside this name. (Eriugena, *Periphyseon*, 1.441A; Sheldon-Williams transl., 25)

Two points immediately jump out. First, Eriugena's *natura*, showing an unrelenting unitary drive, departs from the model of causality that traditionally characterizes Christian creation discourse. While Eriugena divides *natura* into *that which is* and *that which is not*, he never properly defines it. Rather, he expands nature's confines to include even God, as nature comprises not only *what is* but also *what is not*, that is, that which cannot be understood, with negation in a Dionysian vein marking transcendence rather than privation. The inclusion of the divine or nonbeing *within nature*, for only God can truly be said not to be, may suggest Eriugena to be an even more radical thinker than Emerson. For while Emerson, hailed as the philosopher of moods, defines nature as the "not-me"—even if he overlays it at other times with a divine aura, as when calling a farm a mute gospel[68]—for Eriugena, who not just integrates but truly collapses immanence and transcendence, *natura* is God and not-God all at once, which alters the conventional notion of pantheism beyond recognition.

Second, *natura*'s division appears to set up the mind as nature's virtual architect. But does it really? Here a key difference between mind and self as outer tool and inner compass, respectively, comes to light in my reading of the opening line. I read the *Periphyseon*'s famous opening gambit, "Often when I ponder," as embedding the entire work, which is itself a rational investigation, in an antecedent moment of self-scrutiny. Clearly, the faculty of mind or reason plays an important diagnostic role throughout the "rational investigation" (*rationabilis inuestigatio*) that follows. Eriugena attributes being and nonbeing to that which can and cannot be comprehended, respectively. Because being and nonbeing together make up the whole of *natura*, and the mind is the declarer and architect of that whole, it would seem accurate to see the mind's role as coconstitutive with the divine. Given that the mind's activity is placed firmly inside intractable nature, however, the mind lacks an external reference point. In the final analysis, therefore, only the pondering self is afforded that kind of outside grip, even if only in the shape of a momentary flash or transient insight: "Often when I ponder." A Ciceronian borrowing that has often gone unnoticed,[69] the phrase demonstrates that Eriugena's pondering self is ultimately the instance that conjures up *natura*, even if its role is immediately overshadowed by the overwhelming scale of what it has conjured.[70] While *natura* typically envelops and absorbs all things into itself—God, creation, and even the human mind—we will see that, at crucial times, the Eriugenian self has the power to resist such absorption, not unlike how it unleashed *natura*'s unfolding in the first place.

To resist being absorbed into nature, the self needs to exercise constant vigilance. Although the *Periphyseon* presents us with an unwieldy universe, in which the waves of *natura* rise and fall in all directions, it never dissolves into chaos. Standard explanations point to the organizational scheme of procession and return, or *exitus* and *reditus*, whose predrawn Neoplatonic furrows regulate nature's unfolding with the steadfast rhythm of contraction and release. The scheme of procession and return directly underlies Eriugena's better-known second division of nature, namely that into four species. Reflecting the movement from and to God, while studiously avoiding the use of subject-object terminology, these species include *natura creans et non creata* ("nature that creates and is not created," that is, God as creator), *natura creans et creata* ("nature that creates and is created," that is, the so-called primordial causes as Eriugena's combined version of Platonic ideas and Augustinian seminal reasons), *natura non creans et creata* ("nature that does not create and is created," that is, all creatures in the spatiotemporal world), and *natura non creans et non creata* ("nature that does not create and is not created," God as the final *telos* of creation).[71] Collapsing rational investigation into self-scrutiny through what is by all accounts a veritable *tour de force*, Eriugena slowly but forcibly bends procession into return as the *Periphyseon* progresses. I submit that such forceful bending can only come about because the inexorable Eriugenian self lies in wait underneath *natura*'s omnipresence, allowing intractable nature to chart its divinizing course while subtly molding and shaping it.

To explain the absence of psychological terminology in Eriugena, let me give a brief historical digression here on the complex status of early medieval selfhood in the religious West. After the fall of the Western Roman Empire and the cultural collapse that followed, the embryonic selfhood of Augustine's *Confessions* became replaced, and subsequently eclipsed, by the Neoplatonic soul of Boethius's *Consolation of Philosophy*. Early in book 4 of the *Periphyseon*, Eriugena dons diagnostic reason with various biblical traits, as the passage below reveals, and thereby seems, at first blush, to reignite Augustinian embers. While Eriugena may indeed take up Adam as the paradigmatic human being, his project is not one of the confession of the soul but of allegorical exegesis, evincing a more formal approach. Eriugena's engagement in self-scrutiny through exegesis rather than confession leads him to identify the Adam of paradise with the human mind (*nous*) and fallen Adam's necessary toil and labor with discursive reason.

Right when Eriugena formally embarks on nature's return to God, which is the topic of book 4, we see reason raise the stakes of the journey in an apparent attempt to gain strength. This lends it a touch of selfhood, even if the language used is neither intimate nor particularly reflexive. Depicting reason at the helm of a ship in turbulent seas, Eriugena remembers fallen Adam's toil and labor, exhibiting exquisite psychological aplomb as he juxtaposes both cases:

> S[tudent]. Let us spread sail, then, and set out to sea. For reason, not inexperienced in these waters, fearing neither the threats of the waves nor divagations, not the Syrtes nor the rocks, shall speed our course: indeed she finds it sweeter to exercise her skill in the hidden straits of the ocean of Divinity than idly to bask in smooth and open waters, where she cannot display her power. For in the sweat of her brow is she to get her bread (that is, the Word)—so she is commanded—and to till the field of Holy Scripture, prolific as it is of thorns and thistles (that is, a thin crop of interpretations of what is divine), and to follow with the unflagging steps of investigation the study of wisdom, closed to those who spurn it, "until she finds the place for the Lord, the tabernacle of the God of Jacob," that is to say until, the grace of God leading and helping and aiding and moving her by frequent and assiduous study of the Holy Scriptures, she may return and reach again that which in the Fall of the First Man she had lost, the contemplation of Truth; and reaching it she may love it, and loving it she may abide in it, and abiding in it, she may there find rest. (4.744A–B; Sheldon-Williams transl., 383)

While the maritime tropes add an Irish twist to the standard tool kit of the Carolingian liberal arts teacher,[72] (fallen) reason's indulgence in its helmsmanship, its desire to test its navigational skills before arriving at a safe harbor, makes the described quest inalienably Eriugena's own. This is even more the case when he next aligns reason's difficult navigation with Adam's forced labor after being ejected from paradise.[73]

There are a host of reasons for why Adam became the perfect proxy for the early medieval rational self. For one, Eriugena attaches increasing prominence to the Genesis narrative of creation, as he devotes the *Periphyseon's* two final books to the exegesis of the six days of creation, composing a so-called *hexaemeron* that includes the story of humanity's creation and subsequent fall.[74] Long before Eriugena, the tradition of hexaemeral interpretation had forged a collective identification of humanity with Adam (and Eve), laying out the story of the natural world along with that of humanity's fate and history. This parallel between the story of the natural world and that of humanity promoted a close interdependence between macro- and microcosm.[75] The allegorical course of Jewish and Christian thinkers, from Philo to Gregory of Nyssa and Ambrose, showed these exegetical constructs to pivot on Adam's fall in anthropological terms; that is, the allegorization of Genesis allowed them to expound on the paradisiacal Adam as inhabiting an ideal natural environment whose prelapsarian glory would not last. When Augustine replaced the allegorical reading of Genesis with a literal one, much of this exegetical nuance disappeared. Adam remained the prototypical human, but he morphed into a model for the human self that was compromised by original sin.[76]

Following Augustine's fallen Adam, Eriugena presents us with an Adam who is tinged with loss. Melancholy has affected his memory, and the lament for lost innocence that Adam evokes aims to further kindle the longing for paradise of early medieval readers. Many people assume that Augustine's original sin was normative for medieval Christian views on life, namely that it is a vale of tears. An Eriugenian take, however, would be to ascribe life's hardship to the other side of the Fall, which is signaled by God's injunction that Adam get his bread by the sweat of his brow. Interpreting the figure of Adam in terms of early medieval, rather than Augustinian, selfhood, Eriugena opts to embrace Adam's labor rather than his leisure, not unlike how reason prefers rough waters to a smooth sail. It is as if the harshness of Adam's extraparadisiacal effort can be understood, for Eriugena's self, as an opportunity to be seized, a challenge to be met, and, on an intellectual level, a dialectical dilemma about omnipresent nature to think through. The *Periphyseon*'s opening passage fits this same dialectical pattern, with nature absorbing God even though God defies human understanding. As a drawn-out case of Carolingian self-confidence, Eriugena's trek through nature's four species proves more daringly optimistic—if not thereby gnostic or heterodox—about human potential than other premodern religious texts.

Rather than atoning for Adam's fall, Eriugena sets out to complete *natura*'s return in a way that allows fallen reason to safely reach the harbor. Still, even the magisterial heroism with which he bends procession into return cannot undo Adam's sundering from reason's original translucence. What constitutes Eriugena's early medieval genius is that, for him, the Fall opens up a new, albeit painful, avenue of self-reflection in a time and historical context that found avenues only in other enterprises. As a result, he oversees what are, in fact, two simultaneous Herculean efforts, as both *natura* and the reflexive self strive for permanent rest in God.[77] What makes things even more complex, coloring the *Periphyseon* with a touch of the modern and the Emersonian, is that the latter quest impacts, if not conditions, the outcome of the former.

Mystic or Idealist

That rest in God is the goal for both *natura* and the reflexive self raises the question of whether Eriugena is a mystic. As a mystic, he can enter the history of Western religious thought through the backdoor, so to speak, since mysticism does not depend on orthodoxy. Labeling him thus allows us to overcome his condemnation and skirt the pantheism charge. The question of his status has been complicated by the fact that, ever since scholastic discourse came to define the theological paradigm, only a narrow supply of Western medieval authors has qualified as bona fide theologians, with those losing out routinely relegated to the mystical sphere.[78] Augustine may be an overall exception, but Gregory the Great's ascetic-exegetical oeuvre, for instance, read equally widely in the Middle Ages, has been generally considered mystical and ascetic. With the likes of Gregory and Bernard of Clairvaux considered monastic theologians at best,[79] such a label yields its own contradictions when applied to Anselm of Canterbury, who was also a monastic and a theologian. For, although both Schleiermacher and Barth saw Anselm as a forebear to their modern conceptions of faith, he also carries the honorific "father of scholasticism," while others consider him a protoanalytic philosopher.[80] No one calls him a mystic.

Eriugena's fit exceeds even Anselm's in awkwardness. As far as we know, he was not a monk; he precedes the rational mind-set of the schools by a few centuries; and his biblical interpretations, far from an ascetic epiphenomenon, dictate the progress of *natura*'s unfolding with uncanny precision. While calling him a mystic may be the only workable label left, I consider it a stretch even in the nonexperiential sense that the term has acquired today. Bernard McGinn's nuanced judgment of "dialectical mysticism" helpfully invites comparison with other idiosyncratic—one might say self-reflexive—dialectical thinkers like Eckhart,[81] but Eriugena's methodical rational quest cannot but stand in tension to McGinn's rendering of mysticism as the "immediate consciousness of the divine." His observed fondness for intellectual challenge, in particular, makes him reluctant to achieve actual union with God.[82] After all, reaching his goal would cancel out the very challenge on which he so thrives, as the handicap of the course outshines the luster of its completion.[83] Also, it is unclear whether by *reditus* Eriugena aspires to accomplish divine-like ascent or foresees a deeper reabsorption into the intractability of *natura*, of which God is but a part.

If we move from theology to philosophy, recognizing that prior to the twelfth century these areas of study were largely indistinct, a similar uneasiness with fit obtains. In recent years, Eriugena's case has been prominently adduced to support the idea of a premodern, meaning pre-Berkeleian, idealism. Against Myles Burnyeat, who denied that option on the assumption that divine creation in the premodern period inevitably involved the Aristotelian imposition of form on matter, Dermot Moran has built an affirmative case. Basing it in part on the *Periphyseon*,[84] he prominently draws Eriugena ever closer to Gregory of Nyssa and the tradition of Greek Christian Platonism.

Even before the idealism debate with Burnyeat, Moran opined that Eriugena's philosophy "should not be interpreted solely as a hierarchical metaphysics of order, but in fact is an idealist system in which the diversity of nature is understood to be produced by the multiplicity of perspectives of the viewing subject."[85] Rendering Eriugena's connection with the German as well as Berkeleian idealist tradition in nuanced terms, he suggested that the relationship between them is one of difference-in-identity, for which dialectical idealism—like McGinn's dialectical mysticism—might well be an apposite term. For Moran, Eriugena's *natura* unfolds through a dialectical process of divine knowing and unknowing that is duplicated in the mind's creaturely status as divine replica (*imago Dei*, Gn 1:27). This makes the dependence of material creation on intellection crucially important, provoking Moran's claim that idealism was indeed a viable and stable philosophical option in premodernity.[86]

Rather than depending on the imposition of form on matter, Moran regards creation in the Greek patristic tradition as inherently deiform. Creation's theophanic character directly affects human reasoning, to the extent that both the divine and its human image create by self-knowing. As Moran has it, "the paradigm of self-knowing or self-awareness is the founding, thetic, cosmic act. God's self-understanding is the prime mover in the creation of the universe, and in this sense, intellection precedes being."[87] Yet the question is, whose self-knowing is at stake in premodern idealism? Is it the divine intellect that does the self-knowing, or the human, or both? In his deference to the Greek patristic tradition, Moran stops short of declaring creation a product of the human mind, even though his earlier comment appears to see the diversity of Eriugena's *natura* dependent on the viewing subject. Compared to his view that understanding underlies creation's material manifestation, the question of whose understanding, divine or human, may well take second place. Yet it constitutes the crucial difference that separates Eriugena's from Gregory of Nyssa's view of nature.

By detecting an idealist strand that starts with Gregory of Nyssa, proceeds through Eriugena's model of difference-in-identity, and connects up with Berkeley and then Hegel and Heidegger, Moran makes Eriugena a crucial link in a longer philosophical chain. But he does so, I suggest, at the price of tilting Eriugena's Western profile toward Gregory's. Moran writes, "the true essences (*ousiai*) of corporeal things are immaterial and intelligible, and corporeality is simply an appearance produced to the Fallen mind by the concatenation of properties, specifically, quantity, quality, time, and place. Corporeality, then, is a delusion of the Fallen mind, and one which will disappear in the return of all things, when body returns to spirit."[88]

By casting Eriugena as a Christian Platonist à la Nyssa, who considered only the mind created in the image of God and regarded the body to be inferior as image of the image (*imago imaginis*), though Eriugena admittedly takes up this point,[89] Moran overlooks *natura*'s connection to embodiment and physicality inherent in the *Periphyseon*'s larger structure. Similarly, Moran considers the Fall not just as having caused loss, namely, of the theophanic presence of spiritual truth, but also as injecting falseness as the inevitable by-product of the body's shadowy existence. Material creation as diluting spiritual forms of existence may obtain for Gregory, for whom Adam's prelapsarian existence in the allegorical Platonic tradition was more angelic than human[90] (although this point is not uncontested).[91] It does not obtain for Eriugena, however, whose Western realism, rooted in his Augustinian (and Boethian) background, is not so easily denied.[92] St. Paul's motto that "the world in its present form is passing away" (*figura huius mundi praeterit*; 1 Cor. 7:31) certainly holds true for Eriugena, but, following Augustine, Eriugena takes it to reference the form of the world and not nature itself (*figura enim praeterit, non natura*).[93] Accepting *natura*'s intractability as a stumbling block for any permanent demise, as he seems to do here, Eriugena clearly references not disembodied nature alone but *natura* in its fullness of all things that are and are not.

This raises a final issue. In reducing the forms of *natura* to the mind, Moran does not account for Eriugena's self-reflexive moment of *saepe mihi cogitanti* (often when I ponder), which I consider the key frame in which the work is set. This casual phrase, while at first blush bearing out the idealist position, may on closer inspection precisely upend it, contracting Eriugena's so-called idealist project to a mere passing thought.[94] Whether we take Eriugena's gambit as curiously subversive or playfully experimental, with it he enhances *natura*'s intractability to the point that it becomes unclear whether spiritual reality—even if admittedly of a higher order than material reality—should even be considered the pointed goal of all human existence.[95]

To decide this problem, a more thorough discussion of Eriugena's Christology, especially of incarnation, might seem necessary. Maximus the Confessor, whose position Eriugena largely adopts and cites, calls incarnation an act of condescension, a compassionate divine stooping down to humanity's predicament through grace, which, matched by an upward exaltation of humanity through love, effects theophany and theosis.[96] While theophany and ultimately theosis give Maximus his cue, resulting in the self's ascetic imperative to elevate itself ever more with the continuous aid of divine grace, we must analyze further to glean what, if anything, matches this imperative in Eriugena's thought. We will undertake this analysis in the chapter on Augustine and the Postscript that follows. Moving beyond Maximus to the *hexaemeron* of Augustine and, in the Postscript, to his semiotics of giving voice to individual creatures, allows us to chart a different course whereby nature does not only communicate through conversation but becomes itself that conversation.

For now, let me conclude by stating that Eriugena's relationship vis-à-vis idealism cannot but be one of tension. His rational investigation of *natura* is neither primarily mystical nor idealist but, given nature's fundamental intractability, depends instead on his keen antenna for self-examination to stay the course. Where Adam faltered, Eriugena remains firmly at the *Periphyseon*'s helm, or perhaps, in an Eriugenian twist, precisely because Adam faltered, as Adam's failure only solidifies Eriugena's resolve. As the driftwood of Adam's shipwreck rather than the shipwreck itself, the Fall does not only make Eriugena embark on the *Periphyseon*'s cosmic navigation but makes him carry it out until the very end. Far from accepting epistemological privation as defeat, Eriugena turns Adam's loss of contemplation, his ejection from a spiritual paradise that is nevertheless within reach, into a starting-point for the self to begin drawing its circles inside *natura*. They will become widened ever more, as Eriugena pushes outward to attain *natura*'s farthest horizon.

Connecting Eriugena and Emerson on Thinking Nature

Both mysticism and idealism are labels that assume union to be the high point and final destination of Eriugena's premodern quest. In both cases, however, the rush to union can impose a teleological reading that overlooks the subtle ties between literary authorship and religious selfhood. Generally speaking, unless a medieval text is expressly branded as literary—such as the *Roman de la rose* or Dante's *Divina commedia*, both written in the vernacular—it tends to be read for the substance of its (religious or philosophical) content alone. I want to open up a new register of interpretation by bringing in the literary aspect of Eriugena's *Periphyseon*, that of performativity, which can give us better insight into the drama of the self's journey through intractable nature. Approaching the *Periphyseon* as a literary artifact puts its chameleonic hues on full display: it is a Carolingian school dialogue between a master and student.[97] It is a protreptic in which the Eriugenian self tracks the course of meandering nature to find solace and rest in God. And, on a third level, it is a religious or mental exercise to be reenacted time and again.

At the dynamic heart of the *Periphyseon* lies not the idealist question of whether theophany trumps physical creation, since intellection precedes being, but rather the question of what prompts the very possibility of traffic between being and understanding within nature. In other words, what constitutes nature's dynamic? Where does the intractability of nature-in-action, of *natura operans*, find its origin and what sustains it? It is tempting to misread the self-presence that prompts Eriugena to organize the universe *as his own*—that is, as a place where the self can dwell and belong—as an attempt to organize it *on his own*, that is, independent of tradition. To convey their originality, there is no need to lift Eriugena's views out of their historical context, and this in fact does him a disservice. Like Origen, Ambrose, and other patristic authors, he adheres to a biblical script and, also like them, at times departs from it, even if there are points where his sleight of hand is more noticeable. But while the *Periphyseon* features all the requisite themes of God, spiritual beings, the human mind, and earthly bodies, once inside the work, we find that they become parts of *natura*'s interactive play of ever-shifting perspectives, funneled through the four species or modes of the viewing subject. At no point does *natura*'s boundlessness become fragmented or disembodied, insofar as the whole of *natura* shines through each of its forms. Eriugena parades evanescing theophanies but can also zoom in on concretized objects, as when God fashions Adam's body out of clay and plants it in paradise, or when Christ the Word takes on an incarnate body. What binds the theophanic comings and goings and the fleshly bodies of Adam and Christ is that they are equally embedded in intractable *natura*.

Locating the *Periphyseon*'s literary genius in the moment of casual reflection (*saepe mihi cogitanti*) from which nature's arc springs helps us to connect Eriugena in interesting ways with Emerson. It is as if their comparison brings out aspects in both that would otherwise go undetected. Like Eriugena, Emerson is positioned outside the current philosophical canon, whether the continental or Anglo-American one. Unlike Eriugena, however, Emerson's theological exceptionalism is not the product of forcible exclusion but of the awkward public withdrawal from the ministry that marked his early career.[98] Friedrich Nietzsche, who had Emerson's essays translated into German, exhibits a keen susceptibility to his intellectual procedure by referring to them as "attempts" (*Versuche*). By valuing Emerson as a forerunner of his own, much more radical, work, Nietzsche proved remarkably sensitive to the forthright quality of Emersonian language. This forthrightness, standing in apparent tension with his essays' nonchalant or trial-and-error character, has too often been missed by scholars who approach him via the vague prism of his assumed transcendentalism.[99] Although Emerson's forthrightness has been disparaged as a so-called lack of the tragic, I prefer to see it as the hallmark of self-presence, his strategy of "acknowledging" (per Cavell) rather than blindly affirming or reacting to lived reality.

To better frame my discussion of Emerson, I want to deconstruct a popular quotation frequently attributed to him: "What lies behind us and what lies before us are tiny matters compared to what lies within us."[100] As much as this aphorism befits Emerson's style, evident also in his inscription of the word *Whim* above the entrance to his study,[101] it is not he who penned it but Henry Stanley Haskins, a controversial securities trader, who published the anonymous collection *Meditations in Wall Street* (1940). Being off the Emersonian mark, what the phrase gives us is not a dynamic turning of the universe, from outside to inside and back, but a static inwardness that, absent a comparative referent and thus lacking the dynamism necessary for interiority, cannot truly be called interior. Even if we set aside Emerson's avowed disregard for history as "all that lies behind us," the dismissal of "all that lies *before* us" seems decidedly un-Emersonian, insofar as Emerson would not use turning inward as an ostrich-like strategy to close oneself off from reality, least of all from the future. Just as for interiority to resonate, there needs to be a corresponding outside world, so for the present to reverberate, there needs to be an openness not only to the past but to the future. Emerson captures this under the rubric of "onwardness," indicating an injection of temporality that prevents his circles from ever becoming cyclical iterations. As long as the Emersonian self is rooted in nature, it must inevitably move "onward," as nature imparts to it a sacred duty to create, think, and imagine itself on a horizon, that is, in light of a future.

Hovering above the abyss between self and natural environment, as it were, Emerson's aphoristic language can be experienced as disorienting. Thus, in the famous statement, "I am a transparent eyeball," from his first book, *Nature*, the impression of omnipresent nature is so strong (as it is in Eriugena) that one easily overlooks that it is the observing self that mediates the connection between the eyeball and natural reality. When the reader sees the "I" as the mediating, observing self that it is, the transparency of the eyeball, which provides a window on the world, becomes rather more opaque. Although its inherent malleability allows Emerson to push nature outward, away from the self, from the "me" to the "not-me," nature remains closely tethered to the mediating, observing self. Thus, it is unclear where Emerson's observation of nature is just that, an observation, almost scientifically so, through a "transparent eyeball," and where it is an expression or product of a self that has been plied and fashioned by nature.

In terms of the latter, it is hard to overestimate the influence of Henry David Thoreau on Emerson. Thoreau's withdrawal into nature and subsequent testimonial in *Walden* proved an enormous impetus for the circle of New England transcendentalists and brought about a new sense of self-awareness in which his interaction with the natural environment stood for a new accountability of and to the self.[102] Emerson was influenced by *Walden*'s intertwining of what we might call nature-experience and testimonial, the latter reflecting Thoreau's suggested live interaction with his wide literary audience. Emerson thus takes to internalizing the profile of an essayist who is at pains to keep nature and self on a spectrum, but the more passionate he becomes, the more he struggles to tell them apart. No sooner does Emerson manage to carve out a space for the self vis-à-vis natural reality than a discontent with fragmentation pulls him back on the path toward reintegration, forcing him again to blur the line between "me" and "not-me." For example, while the statement "I am a transparent eyeball" shows the self dissolving into nature, it also shows nature contracting to the scope of the observing self. Thus, lodged in the structure of the Emersonian essay is the perennial possibility that the relationship of nature and self may need to be inverted when a different perspective arises. As one fragmented vision gets overturned in favor of another, we see Emerson moving from one "uni-verse" to the next. It is his unrelenting drive for moral perfection that provides movement—and thereby a certain perpetual flexibility and permanence—to these turnings, as his "mood" colors further whatever nature's teaching agrees to impart.

Thus, Emerson presents us not only with an inward-outward dynamic rooted in the circle of natural environment, but by establishing a connection between interiority and onwardness, he also lends his essays a particular forthrightness, compelling us to follow his readiness to confront and think through reality. With the self anchored in nature, furthermore, nature itself has now become temporalized, as it needs to be mediated, proclaimed, prophesied, and above all, lived. Self-presence, by which I refer to the self's rootedness in nature, and self-fashioning, meaning the self's attempts to move outward through nature, are closely connected. Thus, neither pure tragedy nor pure comedy can occur, for such polarized abstractions occlude the self's task of mediating nature's impact wholesale and raw. Instead, we find tragedy woven through the very quality of Emerson's essays, insofar as they are "attempts" that by their very ephemeralness may or may not succeed. All Emerson truly wants to shed is that "paltry empiricism"[103]—that is, the kind of human experience whereby the inner self is so overwhelmed by outer events that it hurriedly patches over its groundedness in nature.

Relating Emerson's moral perfectionism more closely to his "moods," I see the latter as projections of "onwardness" onto nature, reflecting how the self cannot push out into nature without at the same time drawing nature (in)to itself. While Eriugena starts the *Periphyseon* from a spark of concentration, by which the self pushes *natura* forward, for Emerson "thinking nature" likewise includes the self but as the inner core of a circle, not the launching pad for an arc. This makes the self's "thinking nature," in Emerson, truly environmental, as it cannot stop drawing circles around itself, drilling itself into nature, which is already there and will always be there. Throughout its drillings, the self must take care to ride the circumference of the expanding circles, as a surfer catching the perfect wave, maintaining balance while moving on from—and creating—one "uni-verse" to the next.

Panchristology and the Liturgical Cosmos
of Maximus the Confessor

Introduction

Following the discussion of Eriugena and Emerson so far, it appears that elusive nature is infinite in terms of its contours inasmuch as it includes the divine, but when considered up close, it is always one step away from instant evaporation in terms of material content. Furthermore, with the self acting as nature's elliptical counterpoint, the perspectival character of the Eriugenian-Emersonian cosmos makes it virtually impossible to pinpoint where nature ends and self begins. Put in this way, the problem with nature-in-action is pliability rather than pantheism. Nature is linked to a human self that is simultaneously an integrated part of nature's whole and also charged with articulating its connection to nature—to which God is internal rather than external. We find that nature is thus eerily subject to the self's affirmation. At the same time, however, in scale and reach nature far transcends the self, which, as I have explained, mirrors itself in nature precisely to become itself. Cosmic nature is also much richer than can be abstractly delineated. With each observation it is created afresh, evoked anew through a double performance, as the self's mirroring itself in nature prompts the affirmation of nature as nonself that follows, and back again. In this perpetual tango the divine acts as a moving center for a cosmos under permanent construction. Nature surrounds us everywhere, yet it remains uncircumscribable.[1]

Obliterating any fixed causal relationship between creator and creation, nature's identity is malleable and therefore nonessentializable. Locked as they are in a mutual embrace, nature and God, in Eriugena and Emerson, offer reciprocal insight into each other's spheres, as a result of which the natural cosmos takes on an expanded profile and a heightened sense of direction. With the self's desire for God increasingly cast in terms of return (the cosmic synonym for salvation), the directionality of the cosmos takes on more urgency. This is in part because nature's dynamic character—irrespective of whether we conceive of it as a circle or an arc—reveals nature to be inherently temporal. Nature consists of a series of ephemeral moments and, at the same time, charges the self with the task of arranging these moments in a way that is satisfactorily organized and intelligible for the self, even if each moment remains subject to nature's dynamic initiative. As I see it, the directionality of the cosmos adds to the heuristic aspect of nature's profile by inviting us to pose a new set of questions. These new questions focus neither on what nature truly is nor on how we can uncover its deepest workings. Rather, as we now turn to Maximus, our questioning will revolve around when nature, that is, its accordion-like structure fully unfolded and made transparent, can be best deployed in service of human redemption.

Pursuing the powerful questions of how, when, and why nature is to be deployed, I want to devote this chapter to the thought of Maximus the Confessor (580–662 CE). Maximus is a major influence on Eriugena even as his cosmos stands in stark contrast to the airy and at times slippery nature of the *Periphyseon*. Marked as it is by thickness and fatness, Maximus's universe is a useful counterweight to any ethereal cosmos. This so-called fatness is directly rooted in his comprehensive Christology, better called a panchristology. Christ's incarnate presence penetrates and fills out the entire cosmos in Maximus, keeping it fully grounded by adding material weight.

Yet through liturgy Maximus sings the praises of the cosmos in such a way that, while retaining its groundedness, it rises all the way upward to God, fanning out from fallen Adam to risen Christ in a return that becomes ascent. In this breathtaking flight the Maximian cosmos assumes the sacerdotal task of establishing and maintaining a secure connection with the divine at all times, thereby vying with that other body of Christ: the church. This brings the synergy of cosmos and church close to the well-known Maximian parallelism of nature and scripture, which we will discuss below.

Maximus's cosmos is not without its tensions. These stem in part from his Origenian background, in which material reality was a second-best option to an exclusively spiritual reality.[2] In reaction Maximus develops an exceedingly rich and variegated sense of nature, viewing it as a full-blown spectrum of the possibilities of createdness. But the rehabilitation of materiality that his view implies comes at a price: maintaining full cosmic integrity imposes on humanity the Atlas-like task of constant cooperation with the incarnate Christ. Only by such constant cooperation can humanity keep nature's material weight, its *pondus*, from caving in on it. This cooperation occurs chiefly through the practice of liturgy, which, embedded in asceticism, sees the monks sustain and lift up fallen nature in a life of constant effort to sustain and reenact Christ's resurrection.[3] This chapter's guiding question is how Maximus's panchristology succeeds, with the aid of liturgical practice, in lifting up nature and keeping it in motion without sliding back into an Origenian demateralized universe. I will address this by analyzing several key Maximian tropes. In reconstructing the Maximian universe below, even if only dimly approximating the original, a question arises of whether Maximus's panchristological cosmos is sustainable without such liturgical support. Considering, as I do, his actual effort to "think nature" to be successful, how translatable is the model he sets for us? Specifically, what happens to nature's Christological character, and hence to nature itself, if the support structures of liturgy and ascetic practice fall away? Two important follow-up questions to be addressed will be whether Maximus's Christological discourse of nature can survive outside the Byzantine sphere and if and how it can be transferred from East to West.[4]

Staying with the idea of Emersonian circles discussed in the previous chapter, and following the intuition of Hans Urs von Balthasar, I regard liturgy as the widest circle to be visibly drawn inside the contours of the Maximian cosmos.[5] To indicate how closely it approximates nature itself, I have marked two Christological points on its circumference where I see liturgy directly impinge on Maximian nature through (1) the focus on fatness and the meaning of Christology and (2) the focus on the balance of nature and scripture at Christ's transfiguration. By pressing on these two points I aim to push the liturgical circle to become fully convergent with that of Maximian nature itself. This will allow us to see, in Emersonian terms but mindful of the directionality that Maximus adds, how a new and more general circle begins to unfold—that is, that of cosmic return: a new circle to add to our roundtable on elusive nature. In the context of that new circle I will address a third important Maximian point: human nature as the workshop of creation.

While for Maximus the circle of cosmic return can only fully unfold in the eschaton, presenting it here proleptically allows us to "recycle" his anthropological and Christological thought into an integrative discourse of nature. Although that proleptic account will not present us with eschatological redemption, it can show us nature as restored. How to appreciate liturgy's contribution to that restoration is where we must turn next.

Maximus's Cosmic Liturgy

Maximus and His Liturgical Interpreters

My first encounter with Maximus the Confessor was as a student through Hans-Joachim Schulz's study *Die byzantinische Liturgie*.[6] While the movement of "nouvelle théologie" had already attuned me to the ways in which early Christianity could reenergize contemporary theological thought, in Schulz's portrait the horizon of Maximus's liturgical vision stretched this ideal to the theological max. By describing how Maximus was influenced by Justinian's triumphant Hagia Sophia, Schulz invoked for me an imperial and ecclesial world that was in every way the opposite of the ravaged post-WWII European Christian culture. Reaching up to the highest dome, Maximus drew for me the contours of a majestic universe that seemed ready at any moment to burst out of them. His expansive cosmos seemed uncontainable in ecclesial liturgy, yet he aspired to be nothing but liturgical.

Over time my view of Maximus became more nuanced, thanks in part to recent advances in critical Maximus studies[7] and a growing appreciation of his role as an expositor of Dionysius.[8] Yet my fascination with what I would like to call his *sensus liturgicus* has remained. I mean with this term to push beyond historical-critical liturgical study, insofar as the latter can help us contextualize Maximus's ideas but does not mobilize them. Seizing on his *sensus liturgicus* to understand how it is through liturgy that Maximus develops and deploys his theological view of the cosmos, how it is liturgy that puts his nature in flight, I am ultimately most intrigued by the dynamic vigor and temporal quality of that cosmic view. Despite the impressive contribution of Balthasar, whose work filters through in Schulz, the dynamic power of the Maximian cosmos is still largely uncharted territory. Pushing the liturgical circle to match that of nature itself, which I consider an effort faithful to both Maximus's liturgy and his cosmology, I aim to get at the deepest impulse that underlies the Maximian view of the cosmos—namely, the idea that everything in the universe is set in motion, and, importantly, continually affected, by Christ's incarnation. Hence, universal goodness can never be derailed by structural evil or weakened by material deficiency.

One might claim that what Henri de Lubac does for Origen,[9] Balthasar does for Maximus. Through his continued insistence on Origen's exegetical qualities, de Lubac pushes back not only against reading Origen as antimaterialist but especially against reading him as antihistorical. Balthasar's singular achievement is to pick up on the liturgical dimension in Maximus, blazing a trail that Schulz would later follow. In elaborating Maximus's so-called cosmic liturgy, however, that which we typically call liturgy—the set of prayers, sacraments, and rituals pulsating as part of the church's inner life—does not receive much treatment. Balthasar's curiosity is piqued instead by Maximus's Christology, whose Chalcedonian nature Balthasar wields as a seismograph to detect the synthesizing orthodox (i.e., nonheterodox) currents that underlie and flow through Maximus's Byzantine thought. The foundational union of church and cosmos having been sealed by Maximus's Chalcedonian Christology, Balthasar explores cosmic liturgy predominantly insofar as it consists in cosmic symphony.

Compared to Balthasar's "la nouvelle théologie"–inspired view, Schulz's liturgical analysis is more historical, more descriptive, and more accessible. Schulz discusses Maximus's liturgical symbolism in connection with the Hagia Sophia in Constantinople, which was built during the reign of Justinian between 532 and 537 CE (preceding Maximus's birth by several decades), and dedicated to Holy Wisdom, the second person of the Trinity, whose feast of the incarnation was celebrated on December 25. In the setting of this majestic dome church, liturgy was meant to evoke the majesty of heavenly proceedings. Even so, Schulz considers Maximian liturgy in actuality to be rather independent of any church structure, as he sees the liturgical importance of ecclesial space beginning to wane in the *Mystagogy*. In a short section devoted to this work, Schulz expresses concern over the freedom of Maximian liturgy. Departing from Dionysius's gradual but univocal symbolism of ascent from the ecclesial to the heavenly realm, Maximus presents the church in chapters 1–5 of *Mystagogy* not only in connection to the divine but also as an image of the sensible world, the human person, of Holy Scripture and even the soul. A bipolarity of heaven and earth thus replaces Dionysius's hierarchical and mediated ascent, and heaven and earth, settled in their identity, can interpenetrate each other with full reciprocity, meaning that the wholeness of their ensemble is detectable in each.

The *Mystagogy* captures the complex pressures of Byzantine, Chalcedonian Christology, according to which Christ's human and divine natures are always united yet not ever mixed.[10] Yet Schulz applies a narrow lens when he zooms in on the bipolarity of heaven and earth as a way for Maximus to expand the Chalcedonian formula in the direction of the neo-Chalcedonianism of the Sixth Ecumenical Council of 681 CE. This council, the third to be held in Constantinople, defined Jesus as having two energies and two wills, which leads Schulz to focus on the correspondence between Christ's natures (*communicatio formarum*). For Schulz, Christ's hypostatic union is the center of the Maximian universe, which is divided and undivided at once, insofar as it radiates Christ's incarnate presence throughout. This is liturgy for Maximus: a universe upheld by its unmediated relation to the incarnation itself. But Schulz uncovers a danger in the freedom of the *Mystagogy*'s many analogies, because they detract from the primacy of sacrament and the authority of the institutional church. He deems the absence of direct references to the Byzantine eucharistic prayer (*anaphora*) particularly troubling, even though that omission could be explained by Maximus's stated intent that the *Mystagogy* not overlap with Dionysius. Schulz goes on to criticize Maximus for failing to devote a separate chapter to the Eucharist. But when quoting *Mystagogy* 21 on this point nearly in full, Schulz oddly leaves out the one sentence found there that pertains to it. As a result, his truncated version of this chapter sends a different liturgical message than Maximus's original. I quote Maximus's chapter here in full, with the part omitted by Schulz in italics:[11]

> The profession of the "One is Holy" and what follows, which is voiced by all the people at the end of the mystical service, represents the gathering and union beyond reason and understanding which will take place between those who have been mystically and wisely initiated by God and the mysterious oneness of the divine simplicity in the incorruptible age of the spiritual world. There they behold the light of the invisible and ineffable glory and become themselves together with the angels on high open to the blessed purity. *After this, as the climax of everything, comes the distribution of the sacrament, which transforms into itself and renders similar to the causal good by grace and participation those who worthily share in it. To them is there lacking nothing of this good (James 1:4) that is possible and attainable for men,* so that they also can be and be called gods by adoption through grace because all of God entirely fills them and leaves no part of them empty of his presence.[12]

In Schulz's truncated version the final part of *Mystagogy* 21—"so that they also can be and be called gods by adoption through grace because all of God entirely fills them and leaves no part of them empty of his presence"— harks back to the ineffable glory of the union with the angels, which the text mentions prior to the omitted eucharistic reference. But if one looks at the chapter *in toto*, the final phrase conveys the direct, almost common-sense, consequence of the eating of the host.

Contra Schulz, then, I see Maximus making a dynamic liturgical rather than a doctrinal Christological point when he considers the distribution of the sacrament a way for earth and heaven, earthly men and heavenly angels, to be properly linked. Schulz may be right to underscore Maximus's eschatological "union beyond reason and understanding which will take place between those who have been mystically and wisely initiated by God and the mysterious oneness of the divine simplicity in the incorruptible age of the spiritual world"; however, without the regularizing effect of the distributed sacrament, "which transforms into itself and renders similar to the causal good by grace and participation those who worthily share in it," his point is unnecessarily schematic and lacks weight.[13] After all, only those who have worthily consumed the sacrament can "be called gods by adoption through grace because all of God entirely fills them and leaves no part of them empty of his presence."

I have dwelt on Schulz's reductive reading of Maximus's *sensus liturgicus* to draw out what I consider a larger flaw in his, as well as Balthasar's, evaluation. The comparison with Dionysius, Maximus's prime influence, may help us to better identify the nature of this flaw, as we try to pry open the internal dynamics of Maximian cosmic-liturgical thought. Schulz comes back to the Eucharist when he cites *Mystagogy* 24:

(about the "One is holy" and what follows)

> By "the One is holy" and what follows, we have the grace and familiarity which unites us to God himself.

(about communion)

> By holy communion of the spotless and life-giving mysteries we are given fellowship and identity with him by participation in likeness, by which man is deemed worthy from man to become God.[14]

Schulz's citation of this chapter prompts him to make a concluding comment, in which he again negatively compares Maximian mystagogy with the mediating triadic angelic and ecclesial hierarchies in Dionysius. Insofar as all hierarchical stages in Maximus, including the bipolarity of heaven and earth, are funneled through and ultimately united in the redeeming Christ, Schulz sees him betraying an Origenian affinity, though one reworked for a seventh-century post-Chalcedonian audience. Still, this does not prevent Schulz from doling out a stinging rebuke: "Truth be told, for the interpretation of the liturgical symbolism Maximus's dyadic scheme often results in a passing over of those intermediate symbolic meanings which, in terminology derived from the threefold conceptual scheme of scholastic sacramental theology, could be indicated as *res et sacramentum*."[15]

Per this comment Schulz's main problem with Maximus is neither his forgoing of gradual Dionysian ascent nor the transference of Christ's two natures to the radical interconnectedness of earth and heaven, anticipating neo-Chalcedonianism, but his supple and free way of doing so. By collapsing *res* (thing) and *sacramentum* (sacrament), Maximus has connected earth and heaven directly and thereby cut out the guardrail of any scholastic (read: static) third and intermediate layer. By attacking what I see as Maximus's freedom of liturgy, Schulz devalues its powerful resonance in the Maximian universe on the disputable ground that it lacks an in-built, formal mechanism for reining in its dynamics. Lacking any outside means of orientation, Schulz can only see Maximus's cosmos as self-referential, its panchristological potential endangered by the pejorative hues of pantheism.

Making the dynamics of a free Maximian liturgy edge closer toward his cosmology, I will next turn to *Ambiguum* 41 to trace how Maximus bears out his *sensus liturgicus* in a kaleidoscopic, post-Origenian vision of nature.

The Liturgical Cosmology of Ambiguum *41*

Considered a key text in Maximus, *Ambiguum* 41 contains the five cosmic divisions that are central to Maximus's discussion of creation. Taking us to the heart of his Christology, these divisions represent Maximus's orthodox modification of Origen's cosmology of procession and return, while ingeniously inserting Gregory of Nyssa's gender division of paradisiacal humanity. For Origen—whose position in the Christian tradition, even when contested, always remained strong owing to his formidable exegetical insights—a material universe, however perfect, can never be God's first creation. In his Platonic vision material creation was at best a remedy for a spiritual universe gone awry. In exegeting Genesis, therefore, the cosmogony of the spiritual universe comes first, for without it material creation lacks a reason for being. Defending the intellectual freedom of spiritual beings on antideterminist grounds, Origen attributes the dissolution of the spiritual universe to a lapse in their discretionary judgment. The demise of the spiritual universe thus inspired God to create a material duplicate whose *raison d'être* was to allow spiritual beings, now embodied, to return to God.

While the possibility of recurrence remains in Origen, the reverse is also true; indeed, the threat of apocatastasis, the universal salvation of all beings including the devil, ranks among Origen's deepest problems, as it indicates a devaluation of materiality.[16] Since Christian thought from Athanasius to the Cappadocians and beyond is patterned on Origen, his cosmological influence remained considerable; however, red flags could be raised every step of the way, potentially invalidating any cosmological conclusions reached, especially after the council of Chalcedon (451 CE) embraced the two natures of Christ. The vulnerability of an Origenian cosmology may be one of the reasons, together with the closing of the Platonic Academy in 529 CE, why Dionysius expressed himself through liturgy, scripture, and mystical ascent rather than by engaging in speculative cosmology. These problems also may, in part, explain Maximus's move to anchor his views in the ecumenical tradition that stretched from Nicea (325 CE) and Constantinople (381 CE) to Chalcedon (451 CE). Whether or not Maximus's view is sanctioned by conciliar orthodoxy, his chosen paradigm uses the unity of Christ's two natures to legitimize the dynamic connection of heaven and earth. By consolidating their integral connection, it keeps cosmological dualism at bay.

But the question arises as to what extent Chalcedonian doctrine also erects boundaries by constricting Maximus's cosmology with the promulgation of orthodoxy. Or, rather, is the formula of Chalcedon a prism through which Maximus envisions the cosmos in a novel manner, breaking through the Origenian dilemma with a freshly creative approach? As my criticism of Schulz reveals, I lean in the latter direction. *Ambiguum* 41 is a good place to start fleshing out the dynamics of Maximus's cosmology, as its fivefold scheme offers us a blueprint of the Maximian universe. In contrast to Balthasar's presentation of Maximus's cosmic liturgy as cosmic symphony, my question is how we can see Maximus's cosmology as fully liturgical without formulaic orthodoxy reining in or curtailing the cosmos's dynamic freedom.

The standard interpretation by which Maximus's divisions unfold corresponds with Balthasar's symphonic view. On this reading the five divisions mark the genesis and existence of all beings, ranging from the *first* division between uncreated and created nature, to the *second* division of created nature between intelligible and sensible nature, to the *third* division of sensible nature between heaven and earth, to the *fourth* division of earth into paradise and the inhabited world, to the *fifth*, and final, division in which man as the workshop of all things is divided into male and female.[17] This fifth and final division marks as much a crisis as a division, laying bare the rift of Adam's sin.[18] This rift can only be healed by Christ, in whom, according to Galatians 3:28, there is neither male nor female. Bringing to resolution the gender division between male and female with a holistic eschatological vision of the nongendered resurrected Christ, Maximus ensures that the other divisions follow suit. We see, through the ripple effect of ascent, that they are likewise overcome and, in turn, that cosmic return draws ever nearer. In the sweep of one upward motion Christ's unification of the sexes heals the division between paradise and the inhabited earth, between heaven and earth, between sensible and intelligible reality, and, finally, between created and uncreated being. As a result, the cosmos does not merely return *to* God but truly becomes one *with* God. Whereas division carries with it the notion of ignorance, the upward lift of integration accords with a posture of heightened self-awareness, the so-called gnostic *scientia* that Eriugena later takes up.[19]

This much is clear: the fact of Christ's simultaneously held two natures serves as the passkey for interpreters like Balthasar and Schulz with which they unlock the Maximian universe. But while for Balthasar, reading Christ through a Chalcedonian lens means turning cosmic liturgy into cosmic, even soteriological, symphony,[20] for Schulz, Maximus's dyadic liturgical structure risks canceling out the sacrosanct mediating role of the church. Wary that overcoming division might *a fortiori* lead to *Aufhebung* (in the double sense of elevation and cancellation), Schulz would have preferred Maximus to include attention to the intermediate level of *res et sacramentum*, which is where Schulz locates Chalcedonian teaching. Without any firm ecclesiastical grip on liturgy's dynamics—which Dionysius's tripartite division into the celestial ranks, the clerical ranks, and sacramental practice still offers—Schulz fears that Maximus's dynamic universe would become unmoored.

Both Balthasar and Schulz overlook the subtle way that Maximus inter-weaves Christology with anthropology. Rather than reading the Maxim-ian cosmos in terms of soteriology or eschatology, liturgy resists any fixed Christology by giving us tentative insight into the eschaton but in anthro-pological terms. Unlike doctrine, whose static nature Emerson marked as "the height to which the waters once rose," liturgy is inherently tailored to be able to accommodate fraught and fallen human nature.[21] Being too human, too unwieldy, too open to novel practices and reforms to be solidified into doctrine, liturgy offers its interpreters endless opportunities for unpacking. Precisely because Maximus employs so many different Christological tropes and motifs, I see Balthasar and Schulz as pushing too far when they seem to want to limit Maximian liturgy to instantiating, and in the process reifying, only one of them—namely, Chalcedonian doctrine. By contrast I see Max-imus aiming for maximum dynamic effect by pushing the liturgical cycle to match up and align better with the cosmos itself, even if he does so with the aid of Chalcedonian insight. The dyadic tension between heaven and earth brings liturgy newly to life for the ecclesial audience, while making it inch ever closer to the vision of a restored, if not yet fully redeemed, nature.

Keen on overcoming Origen's depreciation of material creation, Maxi-mus's focus on incarnation allows him to parallel Christ's divine and human natures in a comprehensive cosmology that no longer requires Genesis to commence with a spiritual preamble, since in it spirit, matter, and move-ment are now fully integrated. Given that there is nothing the matter with matter, Christ can freely take human flesh. Cosmologically, this grants a freedom and a reality to matter that was impossible for Origen. Maximus's dynamic universe pivots on the incarnation as its kinetic principle, thereby ensuring that the created element will always shine through in his Chris-tology, even as the divinity of Christ is the eschatological horizon toward which heaven and earth move. Rooted in incarnation not only as a kinetic principle but also as a galvanizing moment, a cosmic "event," Maximus's liturgical cosmology brims with energy and begins to unfold in remark-able new ways.

To highlight the importance of incarnation as the hinge of liturgical cos-mology, let us look at the concrete difficulty (*ambiguum*) that he needs to solve. Quoting from one of Gregory of Nazianzen's sermons for the feast of the Holy Theophany (in the East) or the Epiphany (in the West), Maxi-mus states that "natures are renewed afresh and God becomes man." This is a crucial passage that, according to Andrew Louth, was taken up in the liturgical tradition of both East and West. Absorbed into the first *sticheron* of the Aposticha sung at Vespers at the end of Christmas Day, it made its way into the Roman Office, where it was included in the antiphon of the Benedictus for the Feast of the Circumcision: *mirabile mysterium declaratur hodie; Innovantur naturae; Deus homo factus est.*[22] Louth's account of the reception history of this *Ambiguum*'s opening phrase helps us to understand Christ's incarnation as a liturgical moment that, from its inception, was meant to have cosmic ramifications by affecting the natural or created order. According to this view the healing of the cosmic divisions is a cosmic-liturgical consequence of the concrete bond of Christ's two natures: a healing not brought about through a static solidification of each of his natures (foreshadowing neo-Chalcedonianism) but rather incited by the incarnation as dynamic act that reaches all the way down and sweeps all the way back up, so to speak, through the entire universe.

Having established the liturgical nature of the *Ambiguum*'s core query, Andrew Louth next seems to distance himself from both liturgy and incarnation when he states that "the other great Maximian theme developed in this *Difficulty*" is the overcoming of the division of being.[23] One must ask whether Louth does not here do violence to Maximus's discourse, enforcing stasis where Maximus emphasizes flow and movement. We saw how, in the vertiginously dynamic *Mystagogy*, Schulz aimed at keeping *thing* and *sacrament* separate (and in themselves static) through calling for the insertion of a mediating level of *res et sacramentum*. Here, Louth severs metaphysical division from incarnational resonance in a similarly scholastic—in the broad sense of artificial—move. Is he right in doing so? In my view the Nazianzen phrase "the natures are renewed afresh" resonates more as a kind of liturgical *anaphora* (eucharistic prayer), aiming at overcoming division, than as a report on a metaphysical state of affairs. Only constant prayerful repetition, geared as it is at participation in and integration with (the body of) Christ, can bring the cosmos around to unity.

After all, if creation—I use the traditional term here from time to time to stay closer to Maximus's text—is an adequate container for what Maximus calls *becoming* (*genesis*), then liturgy is best seen as an anthropological response to be situated at the level of *movement* (*kinesis*), in participation with and anticipation of the state of *rest* (*stasis*), which is fully achieved only in the eschaton. Following a lead by Polycarp Sherwood, who opened up modern critical Maximus studies, Torstein Tollefsen suggests that Maximus implemented the triad *genesis-kinesis-stasis* (becoming-movement-rest) as an explicit correction to Origen's developmental creational model of *stasis-genesis-kinesis* (rest-becoming-movement). This crucial change makes becoming and movement from painful corollaries of fall and materiality into the engine of createdness.[24] I would add that viewing life on earth as being in constant flux does not only counter Origen's more static, pejorative view of material creation; it also keeps the constant pull of incarnation alive. More than a chastened spiritualist, Maximus declares himself the opposite of a hesychast.

A similarly endemic dynamism permeates the supplementary Neoplatonic triad of procession-conversion-rest, with which Maximus modifies the unfolding of Proclean-Dionysian hierarchies.[25] In its Maximian rendition this triadic principle defines the intricate relationship between the *Logos* and the *logoi* of creatures: each *logos* forms a triad within itself, serving alternately as *logos* of being, *logos* of well-being, and *logos* of eternal well-being, as creation tends toward deification.[26]

Seeing Maximus's foregrounding of division ultimately rooted in incarnation means that each division materializes Christ's presence even in the division itself (that is, before it is resolved in the resurrected Christ), which makes movement much more powerful than a side-effect of creation. Corroborating this, Tollefsen has pointed to the three embodiments (*ensōmatōseis*) of Christ highlighted in *Ambiguum* 33—namely, cosmos, scripture, and the historical person of Christ—indicating that the *Logos* was present in the cosmos even before the historical creation.[27] In a metaphysical cosmology, one might be tempted to compartmentalize these embodiments as three different ontological states or processional stages, but in a dynamic liturgical cosmology Christ can identify with each of them at once.[28]

To the extent that the response that is liturgy can be seen as earthly work (*ergon*), it should be firmly drawn within an anthropological ambit. The task of mediating incarnation to creation is then especially assigned to the human person, who is understood as the cosmos's natural bond (*sundesmos*). This mediation, enacted by the human person, is a prelude to the perfection of Christ's mediation to follow:

> This is why man was introduced last among beings—like a kind of natural bond mediating between the universal extremes through his parts, and unifying [1305C] through himself things that by nature are separated

from each other by a great distance—so that, by making of his own division a beginning of the unity which gathers up all things to God their
Author, and proceeding by order and rank through the mean terms, he
might reach the limit of the sublime ascent that comes about through
the union of all things in God, in whom there is no division, completely
shaking off from nature, by means of a supremely dispassionate condition of divine virtue, the property of male and female, which in no way
was linked to the original principle of the divine plan concerning human
generation, so that he might be shown forth as, and become solely a
human being according to the divine plan, not divided by the designa-
tion of male and female (according to the principle by which he formerly
came into being), nor divided into the parts that now appear around him,
[1305D] thanks to the perfect union, as I said, with his own principle, according to which he exists.[29]

Had man in the first try (that is, as Adam rather than Christ) succeeded in
uniting paradise and the inhabited world through his holy life, then—in
a paraphrase of what Maximus tells us next—he would have fashioned a
single earth; he would have made the sensible creation identical and indivisible with itself; and he would have made the whole of it into a single creation. Had he furthermore, through love, been able to unite created nature
with the uncreated, he would have shown them to be one and the same
through grace. But instead of moving around the unmoved in good Platonic fashion, Adam began to move—contrary to nature—around what
was below him, the Origenian impulse here remaining palpable in Maximus. Putting reality on a dangerous path toward annihilation, Adam thus
divided what was united instead of uniting what was divided:

This is why "the natures were innovated," so that, in a paradox beyond nature, the One who is [1308D] completely immobile according to His nature moved immovably, so to speak, around that which by nature is moved,
"and God became man" in order to save lost man [see Lk 9:56, 15:4], and—
after He had united through Himself the natural fissures running through
the general nature of the universe, and had revealed the universal preexisting principles of the parts (through which the union of what is divided
naturally comes about)—to fulfill the great purpose of God the Father, *recapitulating all things both in heaven and on earth, in Himself, in whom they
also had been created* [Col 1:16].[30]

Through the renewal of natures—both reflecting and receiving Christ's stooping to redeem humanity—liturgy as a human response achieves a kind of resetting of creation. It does so in a motion that may seem counterintuitive, since, through the incarnation as itself inherently paradoxical, the burden of sin is lifted and humanity is restored to full integrity all at once rather than being healed only sacramentally or incrementally.

Thus, contrary to Louth, I see no reason to consider the renewal of natures through incarnation in Maximus as separate from the cosmic or metaphysical division of being. And while Schulz was right to tease out the dyadic structure of Maximus's thought, I do not agree that its absence of a *tertium quid* need be assessed as a lack. Rather, the "work" that liturgy does, that is, the active resetting of creation, brings creation and Christ together directly through prayerful practice. In my view, the *Mystagogy* refrains from eucharistic *ekphrasis* (which would draw out an intermediate dimension) so as not to detract from the performative iterability of liturgical action. The cosmic ascent, propelled by incarnation's resetting of creation, elicits its own *anaphora*, so to speak, as dynamically effective as the administering of the ecclesial sacrament. In so doing, this ascent cancels out the need to isolate the Eucharist as an archimedic point. Such is the power of Maximus's mystagogical embrace, by which he connects not only heaven and earth but, in its retinue, anthropology and Christology, liturgical practice and metaphysical *theoria*. The Chalcedonian Christ serves as a catalyst for this mystagogical embrace, but he is neither its formal nor its final cause. Rather, in accordance with the dynamics of liturgy, the dyadic character of Christ's hypostatic union tilts by nature toward both ascent and expansion.

Two Cosmic Points on the Liturgical Circle

Having discussed *Ambiguum* 41 and its divisions of being within a cosmic-liturgical context, I want to further unfold two points located on the circumference of the cosmic-liturgical circle, as I aim for this circle to touch on universal nature itself—namely, the fatness of incarnation and the balance of nature and scripture (in their roles as intermediaries). These two points will help flesh out the dynamics of anthropology and Christology in Maximus, pushing the resonance of liturgy beyond any doctrinal constraints to its cosmic maximum. After these discussions we will be ready to shift the focus to a third, and culminating, point—that is, cosmic return and the dominant role of anthropology over Christology in it.

The Fatness of the Incarnate Christ

The notion of the fatness of Christ in *Ambiguum* 33, a quandary as short as *Ambiguum* 41 is long, stems from the statement in Gregory of Nazianzen's "Oration on the Nativity" that "The Word (*Logos*) becomes thick."[31] Maximus gives three readings of the meaning of thickness here, not unrelated to the three aforementioned embodiments of Christ (the historical person of Christ, cosmos, and scripture). Maximus's readings are, in the order he presents them, (1) that thickness refers to Christ's manifestation in the flesh, which he takes from and shares with humanity, and which allows him to instruct humanity through words and examples in parables that transcend ordinary speech; (2) that he hides himself obliquely as *Logos* in all the *logoi* of creation, being in them in complete fashion without any diminishing of his simplicity; and (3) that the *Logos* becomes thick for the sake of our thick minds, being not only embodied but also expressed in letters and sounds. All these Maximian readings emphasize aspects of Christ's condescension by which he aims to unite us with himself through the Spirit.

When we next move to the longer *Ambiguum* 31, we see Maximus take up another saying from Gregory Nazianzen's "Oration on the Nativity": "The laws of nature are abolished; the world above must be filled. Christ commands this, let us not resist."[32] Maximus gives us four different readings of this saying, of which the second is relevant, but I will first mention the other ones, insofar as they complement the reading I wish to foreground. Maximus's first reading of Gregory's statement links the abolition of the laws of nature to the unusual manner of Christ's conception and birth, which took place without the usual corruption of conception through seed. Thus, Christ becomes human in a way that bypasses the effects of sin, which allows him to renew the laws of the first and truly divine creation and fill the world above (i.e., heaven) with those who are spiritually born in him. One can hear faint echoes of Origen here but without posing any danger to the integrity of material life. With Christ's physicality maintained, the cosmic-liturgical dynamism prevents the universe's dyadic structure from dissolving into dualism.

Maximus's intent is even clearer in the third and fourth readings. In the third reading the law of nature implies that the nature of each individual thing remains inviolate according to its own principle; it cannot move from its assigned position. Christ, however, in a paradox that is natural to himself, subsists beyond such natural principles, remaining unchanged even as he leaves each nature unchanged. Becoming human and divinizing humanity, he suffers no diminishment of perfection. As Maximus concludes: "This is how God abolishes the laws of nature: He engages Himself with nature amid the things of nature in a way beyond nature."[33] The fourth reading focuses on the partial statement, "The world above must be filled," and takes the notion of Christ to be that of the first fruits (Rom 11:16) of our nature in relation to God the Father, since he is the Word. By becoming man, but doing so without sin, Christ also creates the conditions whereby we can become the first fruits of the human race. As the members of Christ's body become united with Christ as the head, they begin to fill the world above.

The second reading is chiefly anthropological in nature but has Christology effectively tucked into it. Speaking about "my Jesus who is God and the sole cause of all [1277A] things,"[34] Maximus recognizes that Jesus speaks in parables and lists three from Luke 15: identifying man (1) as a lost sheep from the flock of a hundred, (2) as the lost drachma (a silver coin) out of the woman's ten drachmae, and (3) as the prodigal son. In each case, man is defined in close conformity with the paradisiacal Adam dear to Maximus's heart: man is a sheep because he needs tending and is a follower; a silver coin because he is shining, has a royal quality, and bears the image of the archetype; and the prodigal son because "he is the inheritor of the Father's good things and equal to the Father in honor through the gift of grace." It is the third parable, that of the prodigal son, that takes a surprising turn—Eriugena will later pick up on this—when Maximus likens the supreme Word to the fatted calf sacrificed on account of the prodigal son's return. The sacrifice of the fatted calf is Christ's distribution of himself to all beings, which underscores the joy that the dyad of sons is complete, just as the found drachma completes the decad and the found sheep the hundred-fold flock. By applying Gregory of Nazianzus's notion of the Word *becoming thick* to the *fattened* calf to be sacrificed at the prodigal son's homecoming, Maximus suggests this sacrifice to consolidate the connection between Christ's incarnation and the salvation of all beings.[35] The serviceability of the fatted calf is striking here, insofar as it reinforces the thickening of incarnation as conveying the force of the Supreme Word in what may seem a mundane and lowly way. No wonder Eriugena gives this Maximian reading a crucial place in the *Periphyseon*'s eschatology.[36] Maximus himself is more subtle, neither excluding an eschatological reading nor ruling out a eucharistic reference. What is clear throughout is his anthropological point that humans were intended naturally to dwell in the Father's company; it is their separation from God that required the intervention of Christ's incarnation and sacrifice.

The Balance of Nature and Scripture

Here we remain with the theme of Christ's manifestation and its guiding questions: in what way, to what effect, and, quite literally, with what *gravitas* does Christ incarnate? To further explore these questions, I shift attention to Maximus's reading of the biblical transfiguration scene, again picked up later by Eriugena.[37] In *Ambiguum* 10.17 Maximus discusses Christ's transfiguration as he climbs the mountain in the company of Peter, John, and James. There, Elijah and Moses appear to him, and his garments become snow-white.[38] In the process of contemplating that event, Maximus interprets the instant whiteness of Christ's garments by attaching a twofold symbolic value to it.

On the one hand, these clothes represent the words of Holy Scripture, the meaning of which the disciples' intellects were able to intuit without any mediation and without the need to rely on liturgy, one might say. This new intuition or intellection was brought about, precisely, by the moment of transfiguration. On the other hand, Maximus considers the bright white garments to be a symbol of creation itself, which the disciples can now directly take in, devoid of the preconceptions associated with sense perception, insofar as their intellects can penetrate the different forms that constitute creation. In both cases, that of scripture and that of nature, the centrality of the Creator Word is emphasized. There is clearly a soteriological aspect involved, as the original meaning and intent of both scripture and nature (as the proper vehicles communicating the Creator Word) had been rendered inaccessible to human insight. This means that labor is involved in discovering the hidden true meanings in scripture and nature for the rest of us, who were not present at the transfiguration and therefore were not gifted with this insight. It also means that there is a latent dynamic within scripture and nature that makes the discovery process more like an "uncovering" of what was already there.

The latent dynamic to which I refer is not the result of any chase for meaning, for we are not dealing with a dynamic of deficiency. Scripture and creation are understood to be galvanizing vehicles of meaning rather than receptive depositories, yes. But even more than that, inside of them runs an inherent thrust or movement that reaches for intelligibility, reaches for the person who has eyes to see and ears to hear. Without interpretation, without the attempt to bring out what it is they transport, scripture and nature for Maximus remain pools of obsolete knowledge, stagnant in their inability to signify. While it is important to realize that the transfiguration is a moment of luminosity, it is not one that thereby constitutes an upending miracle, for that which is now revealed or brought out is already inherent in earthly life. If there is a miracle, it is the miracle of uncovering, of seeing clearly what is, and has always been, there. Thus, the insight born in this liminal moment has a function comparable to how Maximus elsewhere sees pristine paradise—that is, not just as undefiled by the Fall but without the Fall even on the horizon. The transfiguration unveils the potential of creation by leaving its created status fully intact but lifting its perceived limitations.

By juxtaposing scripture and nature (or creation) at the transfiguration, Maximus inevitably draws them closer, not only in terms of their joint origin in the Creator Word, but also with regard to the collective wisdom to be drawn from them. It is a remarkable feat, one whose enormous importance has been overlooked, for Maximus to present nature as a source that is equal in dignity and value to scripture. Capitalizing here on the deeply exegetical structure of Origen's universe, Maximus combats Origen's spiritualist tendencies. Through the juxtaposition of nature and scripture, Maximus affirms a chastened material cosmos, marking a belated vindication of cosmology. *Ambiguum* 10.18's summary of the shared function of scripture and nature rests on the image of garments: "[1129B] proclaiming the ineffable power of the One who has spoken through it, showing that, by virtue of their reciprocal interchange, the one law is identical to the other, so that the written law is potentially identical to the natural law, and the natural law is by its permanent condition identical to the written law. Both laws simultaneously reveal and conceal the same Word: the one through written words and whatever is visible, and the other through ideas and whatever is hidden."[39] It is interesting that Maximus will later say that these garments clothe or cover *fleshes*, the plural indicating multiple inner meanings or deeper understandings.

Maximus then pushes further by embarking on a deeper explanation of Moses and Elijah, the witnesses or saints present at the event, going through different lengthy interpretations and each time giving us a complementary reading of them as a pair. When Moses and Elijah appeared with Jesus on the mountain, they made such an impression that Peter wanted to erect shelters or dwellings for them, only to have them disappear from before his eyes. The appearance followed by disappearance makes this moment of sight or insight transitory in the sense that it resists reification. And indeed, the moment of transfiguration soon passed.

Among the intricate layered readings Maximus achieves, we find the pair of Moses and Elijah standing in variously for knowledge and instruction, for (ascetic) practice and *theoria*, for marriage and celibacy, and for life and death, as well as types of sensible and intelligible creation, respectively. One of his particularly fascinating readings of Moses and Elijah, and one pertinent to my argument, is that they represent, respectively, time and nature.[40] Moses represents time because he was the first to count the days of creation (as the author of Genesis, as early and medieval Christians believed him to be) and was the leader of temporal worship but especially because he is linked to Jesus, who is "the successor of all time and every age." Playing with the category difference between time (*chronos*) and age/eternity (*aeon*), Maximus calls time the age/eternity that is measured by motion, while age/eternity is time deprived of motion. As a transitional figure, Moses represents the impact of time for, although he was on the brink of entering the promised land, he was not permitted to go there. As the contrasting image of nature, Elijah kept nature's principles inviolate in his own nature, devoid as it was of passion. By means of his discretionary judgment he was also able to instruct those who used nature unnaturally, that is, against nature. For Maximus, who hints at Elijah's correction of sexually deviant behavior here, it is part of nature's integrity to want to punish those who are unnatural, as they can lead one down the path of nonbeing. Elijah's role is that of the guard who protects nature's inner integrity against chaos and abuse, which cancel out existence.

I have emphasized the similarity in the way nature and scripture serve as vehicles of meaning in Maximus, drawn together because meaning is not merely transported through them but galvanized, as a result of which they both radiate their own dynamic thrust. But there is another propulsion at work as well. The prevailing theme underlying this second thrust is the dialectic of concealment and revelation mentioned in the passage above, where written law and natural law each have a manifesting and an obfuscating side to them. Thus, a vertical dynamic is combined with the horizontal one, becoming intertwined with it. It is the complex integration of these two dynamics that the pair of Moses (representing time as a kind of linear and divisible motion) and Elijah (representing nature in terms of the depth of inner integrity) brings out exceedingly well. Joined together, linearity and depth begin to unspool, forming their own circle, so to speak, and it is this foundational or antecedent circularity—admittedly only fully operational in the incisive moment of transfiguration—that Maximian cosmology seems to want to keep in play.

On a larger metaphysical scale, the complementarity of time (Moses) and nature (Elijah) echoes the way in which the temporal incision of incarnation becomes linked to Maximus's deep and circular Neoplatonic ontology. In this ontology, the hierarchical strata of being, according to the aforementioned Proclean-Dionysian stages of *monè* (remaining)–*proodos* (procession)–*epistrophè* (conversion or return), never lose their close connection. On the unfolding of this circular ontology the higher level remains ontologically intact (*monè*), even as it generates the lower level (*proodos*). On and through the latter's return (*epistrophè*) the higher level (*monè*) is reshaped, thereby cementing and enriching the connection between the two levels. While in and of itself this metaphysical ontology is cyclical, inasmuch as it presents us with a world coming into being by going around in multiple circles, the temporal incision of incarnation prevents that circularity from having the final say. By injecting the forward thrust of the Divine Word into nature's circular metaphysics, the incarnation mobilizes nature's operations through what is in effect a divine incision in material reality; this is what I earlier called the directionality of Maximus's universe, which through the incarnation tends toward deification. In doing so, the incarnation opens up the possibility that the moment of incision could be dilated such that the entirety of nature may one day be coextensive with it. After all, are not all of the *logoi* going through the stages of being, well-being, and eternal well-being? This represents no miracle in the sense that the workings of ordinary nature become upended, precisely because nature is created to become ultimately deified. By thus securing the ontological link between incarnation and cosmology, Maximus signals to us the possibility—eschatological or not—that nature's dynamics may one day indeed come to represent a recreated paradise.

But how does the transfiguration affect or transform the ontological impact of incarnation on the metaphysical structure of the cosmos? Specifically, going beyond the incision of incarnation with its vivifying effect on nature, what does the event of transfiguration add to the ontological link between incarnation and metaphysics, overshadowed and truncated as we know it can become by the pressures of ordinary earthly life? For Maximus, the transfiguration provides the impetus whereby the dyad of revelation and concealment, which has been a staple of human language and communication since the Fall, can be overcome in a single and more powerful stream of communication. In this way the transfiguration attaches fresh meaning to this and other comparable dyads such as that of heaven and earth, anthropology and cosmology, man and woman.

Maximus is not one to remain spellbound by the dialectic of affirmative and negative theology. For Dionysius, this dialectic, so foundational for his theological speech, allowed him to gain foresight into what a restored creation might look like and sketch it out for us, as he did in Moses's climbing of the mountain in the *Mystical Theology*.[41] Taking a different approach, Maximus, in elucidating the importance of the transfiguration as a liturgical event, returns to Moses and Elijah's joint appearance with Jesus to the disciples on the mountain. He likens them now, in a more internalized dialectic, to providence and judgment,[42] both liminal terms but also critical and deeply discretionary ones.

In the fascinating segment depicted next, Maximus follows the saints in their ascent, leaving the world and corruptible flesh behind, as they prudently cross the chasm between God and humanity. Maximus mentions that Lazarus (here referring to Luke 16:19–31) was forced to cross this chasm in death,[43] while the saints can apparently traverse it in and with Christ even before death. By crossing over proleptically in Christ, Maximus implies that, like Lazarus, the saints can find rest in the bosom of Abraham.[44] This crossing of the saints gives us an insight into what possibilities the transfiguration opens up for us even in the midst of ordinary time. The transfiguration is an inbreaking that lasts only a moment, but it leaves its mark. After the moment of transfiguration itself has passed, the eschatological vision now unfolded leaves its residue on the reality of ordinary time in which we continue to live.

Nature as Mystagogy: Cosmic Return and the Anthropological Task

If we want to assess whether the two cosmic points I marked on the liturgical circle—the fatness of Christ and the balance of scripture and nature—affect it in such a way that the liturgical circle approximates that of nature itself, we must dwell on the effect of the transfiguration. The transfiguration provides the impetus to keep the liturgical circle in motion, pushing it ever closer to eschatological reality. By doing so, it makes the liturgical circle encroach ever more on the cosmos, turning the latter—since procession and return are now also collapsed—into a wheel symbolizing mystagogical embrace as much as cosmic unfolding.

As we have seen, the *Ambigua* (or *aporiae*, "difficulties") are indeed a good text to turn to in this context,[45] and although I have not wanted to sever its dynamics from the more ecclesial *Mystagogy*, it is the *Ambigua* that has brought us to the threshold of cosmic return. Maximus is especially concerned with establishing a cosmology that doesn't run rampant, that doesn't become loosened from the creator, which would destabilize Christ's central role. That being so, the burden of keeping Christ's stabilizing force in place falls squarely on humanity, failed and fraught as it is. Thus, it is with the anthropological task that we must ultimately end this chapter.

In exploring Maximus's mystagogical intent within his cosmology, I want to take up a comment from Paul Blowers's *Drama of the Divine Economy*, a study on early Christian concepts of creation, in which he sees Maximus not just as a cosmic theologian but as one keen on working out cosmic directionality.[46] As mentioned by Blowers, there are striking, even if mostly stand-alone, similarities between Maximus and Augustine (354–431 CE) in their capacity as cosmic theologians. For both, creation is not merely ancillary but plays a foundational role in establishing the return to the Creator. Blowers writes, "Maximus shares Augustine's concern to assure that creatures' *logoi* (*rationes*) are read not just as ontological archetypes or definitions but as the matrices for creatures' moral and spiritual interrelations with one another and with the Creator."[47]

To bring out creation's foundational status, Maximus emphasizes the role of the *tropoi* alongside that of the *logoi*. In the words of Blowers, the *tropoi* are "the mode(s) of a creature's freedom of movement and its moral and ascetical development."[48] The end of *Ambiguum* 10, discussed in the previous section, deals much with these *tropoi*, thereby emphasizing the need for humanity's contributive efforts to achieve cosmic restoration. But what makes Maximus's position unique—not only in comparison with his Origenian legacy but also with that of the Cappadocians, given that his thought is filtered through the dyophysitism of Chalcedonian orthodoxy—is the way in which he makes Christology permeate any and all cosmic development.

In Maximus we truly have a Christ who not only redeems the entire cosmos but inhabits a cosmos that is geared toward him from its very inception. Maximus thereby prevents any privileged soteriological purpose or Athanasian *theosis* from detracting from creation's inherent, already-abiding goodness.[49] While for Maximus it is still true, along Athanasian lines, that creation is patterned on the goal of divinization, which is realized through Christ's incarnation, passion, and resurrection, what he adds is that the whole of creation, of which humanity is a pivotal part, can only find its true fulfillment when it sings God's praise in complete cosmic-liturgical harmony.[50] The threshold for cosmic return is the liminal moment or the interstitial place—for both time and space are involved in this transfigured cosmic cycle—where cosmology comes to liturgical fruition, so to speak, and becomes mystagogy. It is also the time and place where Christology merges with anthropology, for one cannot address, let alone pray to, God other than through Christ. This prayerful task requires that one continue Christ's work by reorienting the *tropoi*. While that is a charge for all of us, pursuant to the way in which liturgy remains nested inside the church's clerical structure, it entails for some in the church an ascetic life of focused concentration for the benefit of a fulfilled creation.

Ambiguum 7 shows the ingenious strategizing in which Maximus is prepared to engage to prevent theological problems of an Origenian nature from arising, even if in terms of wider orientation he remains largely loyal to a primarily Neoplatonic metaphysical pattern of thought. The Origenian problem lurks unmistakably beneath his interpretation of the chosen passage from Gregory of Nazianzus,[51] which sees Maximus argue that, while the cosmos is characterized by motion, God is immutable. All things being continuously on the move, the danger that they ever reach satiety or stasis and drift away is effectively foreclosed. In other words creation is hardwired in such a way that there will never be a situation where creatures/*logoi* can become bored with or somehow stuck in their divine dependence. Divine dependence in creation *means* movement for the *logoi*. Or, as Maximus puts it in *Ambiguum* 7.7: "Therefore no created being which is in motion has yet come to rest, either because it has not yet attained its first and sole cause, to which it owes its existence, or because it does not yet find itself within its ultimate desired end. Therefore it cannot be maintained that a movement of rational beings previously at rest in a primordial unity [1072D] subsequently brought corporeal bodies into being."[52]

While Maximus diligently heeds the general pattern of *exitus* and *reditus*, inside this broad Neoplatonic matrix, he bends it toward a triadic progression. A *triadic* progression makes the simple reversal of procession and return impossible; with a third point acting as threshold, the important Christological takeaway is that creation is reinforced as inherently good.[53] This is the situation in *Ambiguum* 7.10. Maximus distinguishes here among *being* (τὸ εἶναι), *well-being* (τὸ εὖ εἶναι), and *eternal well-being* (ἀεὶ εὖ εἶναι), laying out a series of gradual steps that see a subtle interplay of *logos* and *tropos*:

> If, then, [1073C] rational creatures are created beings, then surely they are subject to motion, since they are moved from their natural beginning in being, toward a voluntary end in well-being. For the end of the motion of things that are moved is to rest within eternal well-being itself, just as their beginning was being itself, which is God, who is the giver of being and the bestower of the grace of well-being, for He is the *beginning and the end* [Apc 21:6]. For from God come both our general power of motion (for He is our beginning), and the particular way that we move toward Him (for He is our end).[54]

It is in the triadic Maximian model of ongoing creational development, in which incarnation has both a cosmic and redemptive function, that humanity assumes a remarkably elevated position. Maximus follows Gregory of Nazianzus in calling the *logoi* of our being "portions of God," to the extent that they exist in, from, and to God. This both echoes and is strikingly different from Augustine, who, at the opening of the *Confessions*, calls the human person "a portion of your creation" (*aliqua portio creaturae tuae*), emphasizing more dependence than affinity.[55] Maximus addresses this expression in *Ambiguum* 7.20, quoted below. It is important to recognize that leading up to this passage, Maximus has decided to put aside the highest form of negative theology, derived from Dionysius, whereby the *Logos*, because it transcends everything, cannot be understood. Thus Maximus speaks deliberately in the affirmative:

> According to the creative and sustaining procession of the One to individual beings, which is befitting of divine goodness, the One is many. According to the revertive, inductive, and providential return of the many to the One—as if to an all-powerful point of origin, or to the center of a circle precontaining the beginnings of the radii originating from it—insofar as the One gathers everything together, the many are One. We are, then, and

are called "portions of God" because of the *logoi* of our being that exist eternally in God. Moreover, we are said to have "flowed down from above" because we have failed to move in a manner consistent with the *logos* according to which we were created and which preexists in God.[56]

While we indeed have a circle here, with radii connecting the *logoi* to the one *Logos*, we do not yet have a circle in motion. It is as if only the Fall, that burdensome treadmill integrating life as a misdirected cascade of *faux pas*, truly puts the circle to work, injecting the need to transform metaphysics into liturgy and add moral direction, "since we have failed to move in a manner consistent with the *logos* according to which we were created and which preexists in God."

Since Maximus hints here at anthropology as a counterpoint and complement to Christology, I want to leave *Ambiguum* 7 behind precisely at the point of the Fall and return once more to *Ambiguum* 41, where we find that grand cosmic scheme of division and return. Fascinatingly, it is not creation in and of itself that pushes the schema of division and return toward derailment, since movement for Maximus is in essence a good thing, but rather the *synergy* of creation *and Fall*. For as we have seen, the fifth and most crucial division is that by which man—the workshop of all things (*ergastèrion*, or *officina omnium* in Eriugena's Latin), the natural mediator of all these divisions—finds himself divided into male and female. Maximus's interpretation of sin is the opposite of Athanasian *theosis*, inasmuch as sinful humanity contemplates that which is below rather than the God above, along the lines of the aforementioned unnaturalness combated by Elijah. For Maximus, Adam incurs blame for dividing what is united instead of uniting what is divided. Because the ultimate and furthest reach of all these divisions has nested itself in the division into male and female, it is with the single, sexless ("neither male nor female") person of the resurrected Christ that the return via unification of division must start.

To look at the way in which incarnation recapitulates creation, I want to revisit *Ambiguum* 41.6, where Maximus discusses God's intervention through the measure of Christ's incarnation:

> This was why "the natures were innovated," so that, in a paradox beyond nature, the One who is [1308D] completely immobile according to his nature is moved, "and God became man" in order to save lost man [see Lk

9:56, 15:4], and—after He had united through Himself the natural fissures running through the general nature of the universe, and had revealed the universal preexisting principles of the parts (through which the union of what is divided naturally comes about)—to fulfill the great purpose of God the Father, recapitulating all things, both in heaven and earth [Eph 1:10], in Himself, in whom they also had been created [Col 1:16].[57]

For one standing on the threshold of cosmic return, the perspective that unfolds here is that of a cosmic integration of divided natures. Owing to incarnation's holy alliance of God and humanity, as manifest in the two natures of the Chalcedonian Christ, the certain downward road to annihilation brought on by the Fall has by his resurrection been turned into a new upward path of continuous rejuvenation.

But more is at stake in calling Maximus's cosmology a proper mystagogy of nature. For the cosmic rejuvenation to last, humanity is encouraged to mobilize its free will to keep its *tropoi* ("the mode[s] of a creature's freedom of movement and its moral and ascetical development," à la Blowers) from following the certain path of degradation and rather direct them into living a life in conformity with Christ, the *Logos*, who is the archetypal model for all human *logoi*. In fact, humanity is required to do so. The transfiguration gives humans a preview of how the moment of incarnation can be rolled out into a place of redemption, and—through the ancestral scriptural witnesses of Moses and Elijah and the saints modeled on them—shows us the kind of saintly humans who inhabit that space. At the same time, the event of the transfiguration does not move the dial such that we are brought closer to that future time, as Maximus keeps both nostalgic illusion and eschatological utopia at bay. Instead, what the transfiguration does is to warm us to the possibility of its reality. For it to become actualized, however, much more, if not almost all, depends on human nature itself.

Insofar as the church does not expect all of humanity to be up to the task, Maximus charges religious specialists with the obligations of the ascetic life, conferring on them the task of continuous prayer, which culminates in cosmic contemplation. Indeed, cosmic return can only be effected when Christ's cosmological redemption is continued in the human response of liturgical prayer. But to the extent that in Maximus's Christological universe the resurrected, sexless Christ is the axis around which the cosmos revolves, contemplation (*theoria*), at the highest level as the culmination and fruit of liturgical celebration, is assumed to be only attainable in the male ascetic life. As we will see, Augustine and Maximus differ greatly here on the nature of humanity and the human role in the grand narrative of nature, yielding markedly different notions of selfhood.

Conclusion

In conclusion I offer some remarks about how the cosmic return in Maximus unfolds as a mystagogical project in the life of prayer and contemplation. Focused as these are on the division of humanity, my comments will inevitably touch on the inherent putting together of gender and contemplation, embodiment and eschatology.

It seems that, unlike Augustine, Maximus regards sexual difference in the end as something inconsequential, a viewpoint in which he is followed by Eriugena. When describing Christ in *Ambiguum* 41.7, he harks back to humanity's origins, for which he brings in Gregory of Nyssa's famous division of the sexes:

> In this way, He showed, I think, that there was perhaps another mode, foreknown by God, for the multiplication of human beings, had the first human being kept the commandment and not cast himself down the level of irrational animals by misusing the mode of his proper powers [see Ps 48(49):12]—and so He drove out from nature the difference and division into male and female, a difference, as I have said, which He in no way needed in order to become man, and without which existence would perhaps have been possible. There is no need for this division to last *perpetually* [see Heb 7:3], *for in Christ Jesus*, says the divine [1309B] apostle, *there is neither male nor female* [Gal 3:28].[58]

Leaving aside the acrobatics involved in another mode of procreation, what I find most striking in this passage is that in "thinking nature," particularly as far as not just the restoration but the redemption of its integrity is concerned, it is more pressing for Maximus that sexual division be overcome than that human flesh be shed. Be that as it may, he thereby tilts the understanding of the human person toward the masculine, while leaving gender difference hanging in the balance. Obviously, the combination of physical nature, on one hand, and gender, on the other, constitutes a far knottier problem than this book can solve. But that sex, and in its wake gender, remains an obstacle for "thinking redeemed nature," even if human body and flesh are not, is a point worth pondering.

Although for both Augustine and Maximus we can weigh cosmos and prayer on a mystagogical scale, the tipping point for each is different. For Augustine, as we will soon see in more detail, there is less plasticity of created nature, more of an acceptance that the nature one has received and the creation by which one is surrounded is the nature that was there from the beginning. For Augustine this persisting original creation includes sexual difference—even though there is room for debate about whether the existence of sexual difference as part of the original, unfallen creation thereby necessitates gender inequality. For Maximus, by contrast, human nature in its current form is a divided one because of the Fall. Only through the resurrected Christ, whose two natures fully link God and humanity, and whose unsexed unity offers us an integral path to a redeemed creation, can our *logoi* once again be directly linked to the *Logos*. Yet in treating Christ's resurrection as a template, even if he sees Christ as having fully overcome humanity's sexual division, Maximus identifies this division itself as a persistent deficiency. Even though he is intent on overcoming sexual division, since this only happens in the eschaton, and since his sexless Christ in fact remains male, he risks the preservation of sex and gender difference.

As we have seen, it is the synergy of creation and Fall that pushes the Maximian universe on the path to derailment. While the event of incarnation, marking a crucial temporal incision, galvanizes the unity of Christ's natures in a countermove toward restoration, in the absence of eschatological redemption, the Fall needs constant redressing. Even after the galvanizing incision of incarnation, therefore, one needs to continuously secure the conversion of the *tropoi* to prevent distraction and actively remedy the effects of the Fall (death, illness, etc.), which can only be done through a concentrated focus on the resurrected Christ. In this process sexual division turns out to play a precariously troubling role for Maximus. Insofar as it reflects humanity's divided state as tinged with the notion of fallenness, it is a source for lament. At the same time, it gives us a vision and a way forward. The vision is one of how, in and through Christ, the five divisions, including between male and female, will indeed be overcome when the final score of eschatology and embodiment is settled. The way forward is that of reorienting the *tropoi* according to this vision.[59] Inasmuch as this is done inside the church, sexual asceticism is not just a mandatory but an integral part.

In the end the flexibility of Maximus's liturgical focus may make him more vulnerable to gender criticism than Augustine or Eriugena.[60] With

Christian liturgy as an almost theurgical practice that requires the ascetic life to "bridle the passions" (considered a precondition for theurgical practice), Maximus resorts to contemplation or *theoria* as the only way to surpass the push and pull of the passions. Yet contemplation, even if Christocentric, cannot but leave us with a double vision, as the view of how things "actually are," as seen through intuition or intellection, acts as a perennial wake-up call for how things once again ought to be. The more salient follow-up question is whether contemplation, as the culmination point of liturgical efforts, leaves us with a viable cosmos in the meantime. That is, does liturgical glorification based on Christ's incarnation and resurrection not just vicariously replace, and thereby displace, rather than actually heal, creation's broken nature? In consequence, keeping up the artificiality of ascetic life, which presumably sustains the viability of this displacement, underscores the view that sexuality is something to be kept at bay, since in the resurrected Christ sexual division has already been overcome. But since Maximus's cosmos is panchristological, the sexless Christ is everywhere. This renders not only the ascetic life sexless but all of life and all of the cosmos, which in my reading is theologically deeply problematic. The division of male and female is both the last of the divisions to occur and the first to be overcome, making its resolution in Christ the central hinge on which Maximian nature turns. This makes the problem, in other words, one that is not caused by any balancing act of Christ's two natures but by the immovable superiority of his one sex.

While the cosmic-liturgical wheel in Maximus is actively spinning as long as a contingent of male orthodox monks is battling their *tropoi* to conform their *logoi* to the *Logos* in which they have their ultimate place, the separateness and "purity" that this presupposes cannot but cast an eerie veneer over Maximus's cosmology. One gets the sense that nature is forever hamstrung by its inability to fulfill what it was created to do—that is, to effect a return to God through Christ—because it is never separate, "pure," or angelic enough to do so. While the vision of transfiguration fully rolls out the potential inherent in incarnation, the glimpse of eschatological reality cannot be sustained without this constant effort. Thus, what may ultimately be lacking is the awareness of nature as a productive and readily available source for movement rather than a mere site for it. That is, what may be lacking in Maximus is a place that a hurt and sexually divided humanity can inhabit just as fully as Christ inhabits his two natures.

3

Creation and the Hexaemeron *in Augustine*

Introduction
Moving from East to West, our analysis must shift accordingly to reflect the thought of Augustine of Hippo (354–430 CE). Though preceding
Maximus (580–662 CE), in my view Augustine is better understood on
the themes of nature and self when treated after him, because his universe
is best seen as a de-allegorized and de-ritualized one and thus as natural,
material, and standing on its own. This Augustinian universe, which takes
up its own place, is fully embodied and separate from any human need to
make it mean anything beyond this embodiment. This is what I call nature taking place. As we will see, Augustine's creation (the term he prefers
over cosmos or nature) and its attendant terms form a complex cluster of
their own in which the incision of incarnation features no less prominently
than in Maximus's cosmos, though it does so differently. Rather than attributing their difference to the indomitability of Augustinian grace at the
expense of nature, a standard argument, I attribute it to distinct intellectual postures toward nature itself. While for Maximus the rhythm of life is
set in a hymnodic, prayerful key, Augustine combines liturgical and secular life into the reality of ordinary existence. What remains true for both
Augustine and Maximus is that nature, though placed under a differently
conceived aegis in each, unfurls its own course. Far from being simply the
inveterate opponent of Pauline grace,[1] Augustine's nature is a spiderweb of
fine threads that weave subtle and unique connections between God and
self. In what may be the most conspicuous difference with Maximus, the

onwardness of temporality rather than any sense of hierarchy defines the gossamer of nature's web.

The core difficulty plaguing Augustine's deritualized, yet equally dynamic, nature is no doubt its distance from the divine, to which the term *creation* testifies and which its presumed opposition to grace further enhances. As the object of God's creative act, Augustine's nature is at constant risk of objectification, a threat that is held effectively at bay in Maximus's cosmos because of its liturgical flow, in which the divine stooping that is incarnation meets with the prayerful response of the agential human being. Augustine's embrace of material creation is problematic in this regard, as if, by demanding a place for itself separate from the divine creator, creation risks leading to fixation. His embrace of biological procreation as the sole purpose of marriage, having provoked stinging criticism from feminists and other cultural critics alike, further exacerbates the impression of a crude naturalism tending less and less to the spiritual and the grace-filled.[2]

While such criticism must surely be addressed, treating Augustine after Maximus gives us the freedom to approach things from a different perspective. By mitigating the superiority of the ascetic life and rehabilitating marriage, Augustine's approach is not one of reconciling humanity to bodily life as a second-best option but rather of placing the human being squarely in the arena of politics and history. In *The City of God* he presents the idea of life as pilgrimage, extending the meaning of incarnation as much to political as to ecclesiastical citizenship.[3] Thus, he recalibrates the role of the human self—included in, but not defined by, ecclesiastical structure—as first and foremost a sojourner whose life displays significant horizontal thrust.

This makes clear that the Augustinian self is always a temporal self.[4] Without liturgy (ascetic liturgy especially) defining the essential rhythm of cosmic life, freedom and individual choice take center stage in Augustine's nature. The premise of human agency, it seems, is that it operates to move us forward, even as we are impacted by what I will call nature's "taking place," by which I refer to the vagaries of creation's own course as relatively independent from the divine.[5]

Even the most careful extraction of nature, conceived in the dynamic way my earlier chapters set up, from Augustine's wider thought can easily give rise to misunderstandings. Thus, further obstacles will need to be cleared before we can relate Augustinian creation in detailed fashion to the self's forward motion. These obstacles include the opposition of nature and grace, as mentioned, and the Platonic contrast between natural (material, corporeal) and spiritual (intellectual, divine) realities, both of which are present as threads in Augustine's spiderweb of creation. The reason that naturalism—and one may include here embodiment, especially insofar as it extends to sexuality—is a special burden to Augustine cannot solely be explained by these tensions but may well be attributed in large part to what I would like to call his restlessness. Rather than being exclusively definable in psychological terms, this restlessness covers each and every manifestation of created life for Augustine. Thus, if, on the one hand, Augustine can be called the champion of desire, whether he is focusing on the desire for God or using the term in a broader sense, which places focus on the forward movement of the human being, on the other, he often expresses frustration at being entangled in the opposing movements of memory and expectation, pulling him in different directions and thereby rocking him forward and backward. The latter tension projecting what I see as an arc of retardation, a kind of endemic inertia, these movements somehow never quite congeal but slowly undermine the self's thrust as it tries to move forward. In the process Augustinian creation comes to be seen as a weight and an obstacle.

Theologically, it is tempting to attribute the retardation resulting from the self's restlessness—its slow moving forward and backward without delineated order and without the self securing a firm grasp on itself—to the derailing effect of sin. Indeed, sin and evil are more undermining in Augustine than in Maximus, who, not unlike Augustine's predecessor Ambrose, placed more trust in the stabilizing powers of ecclesiastical liturgy than Augustine ever would. Moreover, however we want to assess Augustine's extension of sin to original sin, the latter certainly cast a long shadow over the theological tradition after him, causing him to be blamed for weakening the integrity of human free will and corporeal existence alike.[6] Through their paradigmatic misstep, Adam and Eve seem to have contaminated all of creation in one sweep. To take a fresh look at creation and corporeality in Augustine as the implications of "nature taking place" is all the more important, therefore, as it allows us to pare down the oversized proportions of original sin. Rather than attributing the effect of original sin to the moral motif of *distentio animi*[7]—which locates sin's main effect in the torn nature of the soul and thereby tends to overtheologize it—I prefer to employ phenomenological notions like restlessness, temporal unease, and disorientation to help us better delineate the precarious and specifically Augustinian connections among nature, time, and the self. Far from taking sin out of the picture, this enables us to better track its subtle encroachment on these connections.

There is a further reason to move away from the consideration of original sin as the stereotypical culprit in Augustine. As I have mentioned, sin damages the originary temporal relationship between nature and self through its injection of retardation, causing created reality to be seen as a weight and a burden. Sin undermines humanity's ability to move forward, thereby throwing its future in the balance. This problem becomes shaded differently, however, if we approach original sin not as a derailing device but as a prism. Instead of an obstacle to be removed, original sin sets the conditions for Augustine's rewiring of the triad of nature, self, and time into a fragile and never quite stable order, one that, on the right cultivation, can perhaps point a way for humanity to return to the divine.

As one part of the above triad, nature clearly plays a key role in all of this. No longer a burden, creation is invited to assume its own place, even if that place is within the context of Augustinian restlessness. It is to this process that I refer as creation's taking place. Insofar as Augustine considers the account of creation in Genesis's hexaemeron a case of "nature taking place," I submit that the mere facticity of nature propels him to interpret nature in such a way that he can stake out a new horizon for the human self. In this thereness of nature, this weight granted to its facticity, and by extension opening up a new horizon for human selfhood, he sidesteps both ecclesial liturgy (which might instrumentalize the ascetic human self) and Platonic cosmology (which might seek to overspiritualize creation). Moving from experiencing temporal events as jolts of moral lapse to seeing them as meaningful insights into the intertwining of creation and incarnation, Augustine's vision of nature as "taking place" allows him to fully embrace both materiality and embodiment without relegating either to an inferior position.

For Augustine, a restored connection of time, self, and nature is the remedy for the undermining of both the self's stability and its precarious ability to move forward through temporal unease and disorientation. In order for this restored connection between time, self, and nature to act as remedy for the destabilizing effects of sin, the self must become reattuned to God by traversing a forward-moving path through the created cosmos. A key reason not to assume that the contrast between material reality and divine transcendence in Augustine objectifies the former is that any permanent separation between them would not be able to accommodate the discretionary shifts of movement that such reattuning requires. This forward-moving path, by contrast, ties in directly with time as the prime mover of creation.

It appears that Augustine sharpened his thinking on time—and thereby on shifts in the human being—most pointedly in the Pelagian debate. Throughout this debate the concern to forgo circularity, rooted in Pelagius's idea that Christ's redemption returns us to paradise, makes Augustine increasingly aware that salvation leads us onward rather than backward—in personal life as well as in world history, in church as much as in society. Prior to Augustine the generally held sense was that the removal of Adam's guilt would restore paradisiacal conditions for all, and Pelagius's dramatic radicalization of this notion dispensed not only with the need for grace but especially with the relevance of time and history.[8] By extinguishing Adam's sin almost as if it never happened,[9] Christ's redemption in Pelagius may seem to pave the road to paradise for all, but in reality it secures entry only for those militant ascetics strong enough to avoid repeat offenses. The vast majority of Christians, meanwhile, are tricked into a mirage or *fata morgana*, as paradise sets them adrift in what is, in effect, a postlapsarian nature without goal or meaning. More than anti-Pelagian strategy alone, Augustine's definition of salvation as distinct from the return to paradise bears out his underlying pastoral concern. By cutting through soteriological circularity and dispensing with any mirages,[10] he protects the self from wandering about aimlessly in an ever more disordered universe.[11] His is an attempt to rethink creation in terms of ordinary, sinful life: that murky and imperfect whole that is yet the only reality in which the temporal self will ever find itself. In this same reality the temporal self must find its way back to God.

For both Augustine and Pelagius baptism washes away sin—that part of the Christian heritage they share in common—but for Augustine it does not thereby fully expunge the staining of Adam's originary sin. Ordinary life remains fraught, therefore, as the luster of salvation inevitably fades. While restlessness can make the self indolent—its resolve languishing as it oscillates between expectation and memory—Augustine's acute awareness of temporality gives him the tools to frame life as a series of temporal incisions and movements deeply animated from within. Through predestination and perseverance—predestination as the divine gift of time, and perseverance as the self's affirmation of it—memory and expectation need no longer be considered distractions but can become newly coordinated with the divine.

It is misguided to think that predestination turns the divine into a control freak or that perseverance forces human beings to run their lives as an austere celestial marathon, as if divine governance by definition detracts from the enjoyment of ordinary life. It is equally untrue that God is sitting back watching with relish sin's destructive impact on human life. As much as the self's endemic restlessness precludes any moment of triumphant restoration, death for Augustine is not ever natural or timely. In the newly configured view of time, self, and nature, however, the arc that connects memory and expectation, predestination and perseverance, can now lift up rather than press down the burdensome weight of time.

How to assess Augustine's respect for everyday, individual ordinariness—what Robert Markus has called his embrace of Christian mediocrity,[12] which I see not as a bishop's compromise but as reflecting the self's deeper defiance of resignation in the face of life's ever falling short—is a crucial question. Also crucial is how to evaluate the directional impact of time on the status of creation and createdness, for it is under the same arc of predestination and perseverance that nature takes its place alongside the self. In the next segment I will focus on Augustine's exegetical reflections and their correlation of nature and self.

Scripture's Mediation

One reason that creation and createdness impede easy analysis is that creation in Augustine's oeuvre is such a pervasive theme. Presenting itself even when one does not expect it, its most notable surprise occurrence is no doubt in the *Confessions'* last three books. While Augustine is given to cosmological musings earlier as well, here, creation comes into full view, drawn as it is into the ambit of Augustinian selfhood. While creation's role in the last books of *Confessions* has been duly noted, its corollary—namely that Augustine's careful ordering of the *Confessions* means that creation is fully integrated with the perceived dynamic of the entire work—is generally underestimated. This dynamic is perhaps best summed up in the famous outcry, "I have become a problem to myself."[13]

To forge a connection between creation's pervasiveness and Augustinian selfhood, the apt description of his becoming a problem to himself found in *Conf.* 10.33.50 is revealing. Troubled by the impact of church music, whose seductive effect may distract him from concentrated attention to God, Augustine addresses God and his readers alike with the following lament:

"See my condition! Weep with me and weep for me, you who have within yourselves a concern for the good, the springs from which good actions proceed. Those who do not share this concern will not be moved by these considerations. But you, 'Lord my God, hear, look and see' (Ps 12:4) and 'have mercy and heal me' (Ps 79:15). In your eyes I have become a problem to myself, and that is my sickness." Augustine's famous theme of "I have become a problem to myself" emerges here at a particular moment of self-examination as we find him prostrating himself in biblical fashion before the eyes of God (*in conspectu Dei*), having arrived at an emotional low. With God and the self intimately connected—not just because *Confessions* is a dialogue in need of an addressee but also because since the early *Soliloquies*, God and the soul are all Augustine ever desired to know[14]—creation seems like *der Dritte im Bunde* (the third player), a concept for which there seems to be no specific, guaranteed place. Creation is mostly just there, without a clearly assigned place or well-defined role, neither thematized nor particularly problematized.

Pervasive but undertheorized, creation has a translucent quality in Augustine, causing it to be either overlooked or subsumed under the more central rubrics of God and the self, both of which feature in Augustine's most prominent monographs, *De trinitate* and *Confessions*.[15] As if unable to account for its translucence, scholarship on Augustinian creation often subjects it to standard treatment whereby, amid the usual tally of creatures passing review from animals to angels, attention to Augustine's framing of creation is strikingly absent. This is the case, for example, in Simo Knuuttila's article "Time and Creation in Augustine," where the author's comment early on that "there was nothing radically new in Augustine's conception of the creation" sets the tone. Knuuttila makes a few relevant cosmological and philosophical points, focusing especially on the concept of *creatio ex nihilo*, but thereafter dwells mostly on the notion of psychological time, evoking points not unlike my above connection among time, self, and nature but without analyzing them further.[16] Similarly, in his aforementioned *Drama of the Divine Economy*, Paul Blowers gives us a summary of Augustine's views that predicates creation on familiar ideas of time and evolution.[17]

Inspecting some of Augustine's more salient comments on the topic, with or without explicit ties to the self, may offer us better access to the specifics of Augustinian creation. Right at the beginning of *Confessions* we have an interesting case of creation's thrownness, its simply being there. After first introducing humanity as "a part of your creation" (*aliqua portio creaturae tuae*), Augustine next states that "you, God, have directed us towards you" (*fecisti nos ad te*) and only thereafter, referencing humanity a third time, employs the first-person language that has so come to identify the work.[18] But when we look at Augustine's description of *Confessions* in his *Retractations*— his not-to-be-underestimated annotated autobibliography—the difference is striking. Using first-person language without any qualms here, Augustine boldly states that "the first ten books are written about me" (*de me scripti sunt*), only to continue by stating that "the other three are about the holy scriptures" (*de scripturis sanctis*), by which he means Genesis's hexaemeron "beginning with what is written (*ab eo quod scriptum est*): 'In the beginning God created heaven and earth,' until the Sabbath's rest."[19]

Exhibiting deeper insight than is customary, this brief entry from *Retractations* reveals a subtle but daring parallelism between what Augustine writes about himself and what he writes about the holy scriptures—that is, Genesis. Aided by what I consider a parallelism of texts (about self and about the hexaemeron) rather than objects or themes, Augustine directs creation away from its translucence whereby it is either overlooked or contracted with humanity or God. He thus unlocks an altogether different and deeper dimension, one in which creation and createdness are interwoven with everything Augustine says. This interwovenness does not only affect, in thematic or topical fashion, what Augustine says about the self. If we go straight to the heart of the matter, this interwovenness marks all his speaking and writing—which are, as they are in God, inseparably connected—as fundamentally creative. Thus, not only is it true that when Augustine references himself as *aliqua portio creaturae tuae*, creation is speaking in and through him, but his very speaking (and writing) is itself also to be considered a kind of creating.

The latter point has been teased out by Sabine MacCormack in her 2007 Saint Augustine lecture "Augustine Reads Genesis."[20] When first reading her lecture, I worried that it shortchanged scriptural exegesis, as MacCormack focuses on Augustine's *manner* of reading Genesis rather than on what he actually read there. What, if anything, I wondered, could Augustine's mode of reading Genesis reveal programmatically about his view on creation's materiality? As it turns out, a great deal, especially if we factor in Augustine's statement in *De Genesi ad litteram* 1.15.29, which lies at the basis of MacCormack's analysis. In an attempt to grasp the intertwining of the mode of creation and the matter (the literal "stuff") of creation, Augustine compares divine creating to human speaking. Just as humans do not first emit sound and later fashion words from it, he argues, so God also did not first create a kind of primeval matter from which later to finesse individual beings. He formed and fashioned all things at once (*omnia simul*), preventing any hierarchy between the spiritual and the corporeal from taking root:

> Just as a voice, after all, is the basic material for words, while words are what a voice is formed into, but the speaker does not first give vent to an unformed voice which he can later gather up and form into words, so too God the creator did not first make formless material and later on form it, on second thoughts as it were, into every kind of nature; no, he created formed and fashioned material. If the question were asked, I mean, whether we make a voice from words or words from a voice, it would not be easy to find anyone so slow of wit as to not answer that it is rather words which are made from a voice. So too, although the speaker makes them both simultaneously, it is clear enough, on a moment's reflection, what he makes out of which.[21]

MacCormack's conclusion that "just as human speakers form words out of sound, which is the matter of language, so God the creator formed the visible universe out of inchoate primeval matter,"[22] is hence fully justified. But if we elevate her topical analysis to a programmatic level by deploying the parallelism of speaking and creating, it appears that the issue of materiality—the sound of speech, the matter of creation—may not be what is ultimately at stake for Augustine. Perhaps out of an intuition that the best way to cope with creation's pervasiveness is to break it down for the purpose of understanding, what Augustine appears most keen on detecting through reflecting on creation's matter and mode is a tangible principle of order and organization. Hinting at the notion of order, the above passage from *Gen. litt.* 1.15.29 distinguishes between a priority in time and a priority in origin, with primeval matter pertaining to the latter but not laying claim to the former, since everything is created all at once. Human intellect and speech inevitably fall short in their attempt to express the simultaneity of creation. For this reason Augustine considers scripture an indispensable tool that can lead us out of this impasse. Creation is called into being by God's very Word. Scripture, understood here as God's words, is a divinely sanctioned compensation for humanity's inability to understand and express the all-at-once-ness of the creation of all things. Scripture helps us in this regard by unfolding over time, according to the days of creation.[23]

Thus, Augustine states the following: "It is not because formless matter is prior in time to things formed from it, since they are both created simultaneously together (*utrumque simul concreatum*), both the thing made and what it was made out of; but because that which something is made out of is still prior as its source (*origine*), even if not in time (*tempore*), to what is made from it, that scripture could divide in the time it takes to state them what God did not divide in the time it took to make them."[24] *That scripture could divide in the time it takes to state them what God did not divide in the time it took to make them.* . . . Mark how in this terse comment, which in fact precedes the earlier quotation, Augustine does not go so far as to say that God *could not* have divided things at the time of creation by creating primeval matter before forming individual creatures, only that he *did not* do so. By asserting that scripture *could* separate them out, he emphatically suggests—even though he does not spell it out—that it *did* indeed do so, containing ample wisdom to help us unpack the proper meaning of creation and createdness.

If my interpretation is correct, Augustine's parallelism of the projection of creation and speech, premised, as MacCormack highlighted, on the interrelation of their matter and mode, reveals a deeper underlying asymmetry between them; for creation is God's to fashion and speech humanity's to shape. Affecting both spheres, the problem of human inadequacy looms large, even if in rather different ways: in creation we are faced with a God who acts without stooping to explain himself, aligning creation with predestination; in language humanity lacks the capacity to ever rise above its created status and accurately express the towering majesty of the divine. With divine creation projecting a sense of divine omnipotence, and human speech hamstrung by the inability to do justice to the divine (whether sininduced or not is inconsequential here), Augustine unfolds Genesis's hexaemeron as a scriptural pedagogy that can effectively mediate between these two spheres. While scripture conveys divine omnipotence, it does so in a way that allows for human insufficiency.

Situating scriptural narrative on a mesolevel between God's actual creation and humanity's flawed understanding allows Augustine to bypass (but not erase) their ontological asymmetry. To do so, he seizes on the hexaemeron as a channel, separate from either God or humanity but connected to both, through which the events of creation can be passed truthfully and effectively from their divine givenness into workable human understanding. If Augustine ever embraced any epistemological principle, it would be that of scripture's tertiary nature, by which he straddles divine majesty and human inadequacy in a way that is peculiarly his own.

Augustine's pledged return to God through scripture ultimately means that henceforth the human self can only come closer to the reality of the divine by working in and through creation.

From Platonic Cosmology to Literal Exegesis

Given Augustine's respect for the hexaemeron as a divinely sanctioned pedagogical template on which to predicate humanity's understanding of created reality, his account of creation could not take the shape of a work like the speculative *On First Principles* of Origen of Alexandria (ca. 185–254 CE).[25] It should be noted that Origen's masterpiece, like many masterpieces, has been prone to misunderstanding, which is exacerbated by the fact that this masterful exegete invokes exegesis only in the fourth and final book. Thus, the work has been taken for a Platonic cosmology with an exegetical veneer, or a Middle-Platonic physics with a biblical postscript. Intent on correcting such views, Brian Daley reads its Greek title, *Peri archon*, as deliberately ambiguous, inasmuch as he considers Origen to be keen on remaking *exegesis*, in addition to ancient cosmology, from the very beginning of the work.

In Daley's words, Origen's way of "constructing a cohesive survey of the ontological principles of the world's being . . . also brings together, for him [Origen], the logical principles for an understanding of the content of revelation that is both the anchor and starting-point of authentic and creative biblical interpretation."[26] Using as his guide a then well-known list of what the apostles had taught, in the *Peri archon* Origen draws the contours of an Aristotelian science or *epistèmè* that, on the merits of its intrinsic plausibility, tries to infer the many things that the apostles had left untreated. Far from embracing modern inductive methodology, Origen's science represents the growing organization of a whole body of available Christian data, containing scriptural references and doctrinal premises, as well as the conclusions drawn from them. All along, Origen's final aim, per Daley, was to authoritatively formulate a Christian *epistèmè*, in which the various interrelations and explanations that could be detected were clearly laid out.

Scriptural input thus impacts this science from the very beginning, and Daley considers Origen's real interest—foreshadowing Augustine at least in this, if not in other regards—to be in what he calls biblical theology. Origen's conviction, however, was that a productive, responsible reading of scripture was truly possible only after developing these first principles as the contours of a science that, until its eschatological completion, would necessarily be embryonic. Awareness of the *Peri archon*'s setup thus helps us to understand why the discussion of biblical hermeneutics proper comes so late in the work. Rather than proving exegesis to be an afterthought, its late treatment brings to the surface what has been scripture's implicit, even if tacit, involvement in the development of Origen's science from the work's very beginning. That is, the entire work is building the conditions by which we will be able to interpret scripture effectively.

The methodological difference between Origen and Augustine gives us insight into why the cosmological account in the former is absent from the latter. Indeed, Augustine never formulates a natural theology outside the bounds of scriptural exegesis. Nor do we find in him something akin to the hybrid format that Eriugena, successor to his creative genius in the early Middle Ages, puts forth in the *Periphyseon*. That work, as we have seen, opens with a dialectical physiology that segues more or less organically into a hexaemeral commentary, integrating the comprehensive theoretical concept of *natura* with the individual natures that populate it. The merger of physiology and exegesis in Eriugena has its roots in Maximus's parallelism of nature and scripture. This parallelism is premised on Maximus's view of the cosmos as a liturgical expansion of incarnation, which leads him to see nature and scripture as the two garments of Christ at his transfiguration. In the same vein we can see the individual creatures of Eriugena's hexaemeron reinforce the credibility of *natura* by literally fleshing out the teaching of scripture.

But what Augustine gives us in *De Genesi ad litteram*,[27] which serves as my text of comparison with *Confessions*, is a fully fledged, well-rounded hexaemeral interpretation. Augustine had access to and knowledge of Basil of Caesarea's first nine *Homilies on the Hexaemeron* in the translation of Eusthatius,[28] who had also been an influence on Ambrose. Although Eusthatius's translation did not include Basil's *Homilies* X and XI, I want to draw attention to what Philip Rousseau concludes from these later sermons about Basil's exegetical purpose in his illuminating article "Human Nature and Its

Material Setting in Basil of Caesarea's Sermons on the Creation."[29] Since Rousseau's analysis elucidates Basil's earlier homilies (the ones known to Augustine), as well as his later ones, his view can shed important light on Augustine's wider purpose. For one thing, Rousseau adamantly rejects the persistent idea—still suggested, even if not highlighted, by Richard Lim— that by foregrounding the literal reading, Basil seems to apply a quasi-Origenian distinction between the educated and the uneducated.[30] The assumption here is that foregrounding the literal reading entails speaking to a broader audience as opposed to an intellectually and spiritually elite one. As Rousseau powerfully retorts: "It is careless to suppose that breadth of address diminishes the quality of Basil's rhetoric."[31] This in itself could be a guiding motto for any serious reader of Augustine as well.

Having cleared up the matter of Basil's supposedly anti-Origenian agenda, as if his literal program in reading Genesis connotes negative valence only, Rousseau explains from the opening passage in sermon 1 that the notion of *text* is always central in Basil's homilies. Basil's notion of text may alternate between scripture's speaking and its writing—varying furthermore between Moses (he) and scripture (it) as the one doing this speaking or writing[32]— but he remains steadfastly convinced throughout of the need for loyalty or fidelity to scripture. Expanding the orbit of these themes, Rousseau touches on two crucial points that are directly relevant for our broader discussion of Augustine: (1) that the reading of scripture underpins the construction of self, as human identity is at root the identity of a hearer and a reader[33] and (2) that with regard for the multidimensional creation of humanity, including the concrete and material, Basil distinguishes between the form of *historia* (ἱστορία, narrative) and the power of *theologia* (θεολογία, theology) as distinct exegetical levels. In elaborating the second point, Rousseau states compellingly that in the matchup with history theology proved in general to be the less limiting endeavor, or perhaps the more inclusive one, insofar as it allows for interpretive movement upward as well as onward or forward. By instructing the reader in the myriad possibilities of how to read, theology cultivates "the infinitely progressive character of the spiritual life" for which the Cappadocians have become well known.[34]

Rousseau steers clear of any misconceived hierarchy of the literal and the spiritual senses being primitive and refined, respectively, as such a hierarchy tends to assume that the literal need not be interpreted. What must precede "right reading" is setting up the right hermeneutical relations between observer and observed, which requires expansive and full attention to the observed, that is, scripture. This means that all meanings, not only the spiritual ones, are worthy of expression.[35] For Rousseau, these hermeneutical preconditions are necessary to get to an accurate observation of Genesis's materiality, which itself becomes worthy of interpretation on its own terms.

The fact that the act of God's will (βουλή, *boulè*) to create precedes his act of creating humanity, which Rousseau interprets with an eye toward humanity's special creaturely dignity as *imago Dei*, does not hierarchize the important distinction between the two Greek terms that are used to describe humanity's creation. These are *poiesis* (ποίησις; creation as *making* humans in God's image, cf. Gn 1:26–27) and *plasis* (πλάσις; creation as *molding* humans from the earth, cf. Gn 2:4). As Rousseau explains Basil's reading, God is working from the inside out, as "God's primal and active skill ordered things at the deepest level, working progressively from within" (ἡ δημιουργικὴ αὐτοῦ ἐνέργεια πάντα διωργάνωσεν ἐπὶ τὸ βάθος χωρήσασα ἔνδοθεν, *Hom.* 11.14).[36] Because of its material nature, *plasis* (molding) would on a Platonic scale be considered a secondary step or mode of creation after humanity's first, "spiritual," creation in God's image. But since it is undertaken by the unchanging, undifferentiated divine, *plasis* is neither an activity that is inherently lower than *poiesis* (making) nor one that lowers human beings. Rather, the opposite is true: the erect posture of humans is a direct product of their being molded, enabling them to look upward to God in accordance with their dignified nature. In what Rousseau calls "a cartography of moral development," this erect posture allows humans not only to carry out the goal of their creation but also to become a lesson that bears out this goal for others. The fact that humans are molded can become an important aid for them in the return (*Hom.* 11.5) to their former natural relationship with God.[37]

If we now return to Augustine,[38] Rousseau's comment that a literal meaning is not inherently lower or less complex deserves special attention. According to Karla Pollmann,[39] elaborating on the anti-Manichean character of *De Genesi ad litteram*, Augustine deemed a literal commentary to be particularly necessary because he considered his earlier *De Genesi contra Manichaeos*, written after he had recently resettled in Africa, still too allegorical. And, since the Manicheans despised allegory, it was thus inadequate to settle his dispute with them. Pollmann explains how the term *literal* in Augustine's *De Genesi ad litteram* is not reductive but displays a wide variety of meanings. It can refer to what really happened (*Gen. litt.* 1.1.1), which is the conventional sense of *literal*. But since it is specific to Genesis that things happen for the first time, the literal can at times be referring to a spiritual reality. In addition, the literal as a "first sense" of the *letter* of the text can sometimes be spiritual (*Gen. litt.* 8.1.2), as for Augustine the text of the Bible itself can sometimes have a figurative meaning (*Gen. litt.* 8.1.1). The literal sense can also refer to the truest meaning (*Gen. litt.* 4.28.45),[40] or it can indicate what is obvious to everyone based on sense evidence (*Gen. litt.* 2.9.22, *omnium sensibus patet*). So, *literal* does not necessarily map on to an ontological hierarchy. For Pollmann, Augustine's main difference with Basil and Ambrose is that he does not use nature itself for purposes of moral edification. Instead, he sees nature becoming dangerous only because of human sin, as the thorns created at the beginning become harmful to humans only after the Fall.[41] Thus, Augustine seems to place the blame for any perceived natural consequences of the Fall entirely on humanity.

Reluctant to absorb everything too quickly in the great Augustinian spin-dryer of fall and original sin, his choice for literal interpretation aptly conveying human sinfulness, I deem it prudent to edge closer to Rousseau's account of Basil in explaining *De Genesi ad litteram*. Mindful of Rousseau's subtle call for attention to the setup of the relationship between reader and text, I see this dynamic in Augustine as fundamentally different from the dynamic in Basil. Notably, as a hearer or reader of the biblical text Augustine conceives of himself now also as a writer.[42] As evidenced by his statement in *Retractations* about *Confessions*, Augustine sees a parallel between what he had written "about himself" (*de se*) and what Genesis says. Even if *De Genesi ad litteram* does not follow *Confessions* in all respects, it can be assumed that the art of writing is something that is on his mind in that work also—that is, the art of how Genesis was written, which, in turn, has implications for how he must write about Genesis. Conversely, we can also assume that Augustine's discussion of creation contains an element of self-generation outside *Confessions* as well. It is in this vein that the literal meaning of Genesis cannot but be crucial for him: not unlike how the notion of *imago Dei* according to Jean-Luc Marion identifies the human beings not as copies of God but as "gods,"[43] scripture imparts the signature of God's authorial manifestation directly in and to creation.

In its status as the most fundamental text *qua* drafted and crafted product, the literal level of scripture is the level at which the tone and direction of all communication are set, as meaning goes out from the speaker or writer (God, Moses) to the hearer or reader. Compared to the literal level, all other levels—be they allegorical, spiritual, figurative, or typological—provide us with additional information, whether by way of edification or embellishment or even to mark divergence from a *prima facie* reading. In his commentary, which is expressly *ad litteram* (that is, according to the letter), these additional levels tempt Augustine to veer off into digressions from time to time. But his excursions always end with a return to the literal biblical text.

Creation and Conversion 1

Clearly, then, for Augustine there is no sound reflection on creation or createdness without text—that is, Genesis's hexaemeron—the reading of which should not be muddled by any exposition that is too overtly one's own, because truth is contained in the text itself. Elaborating this point in a seminal article about the role of the hexaemeron in Augustine's reading of creation while putting her own twist on scriptural pedagogy, Marie-Anne Vannier links creation to conversion. To this end she borrows from Gerhart Ladner's *The Idea of Reform* the notion that "creation itself is unthinkable without God's immediate recall of his creatures to Himself and their conversion toward Him."[44] Aware of Ladner's interest in the overarching triad of creation, formation, and reformation, Vannier modified Ladner's construct to reflect her underlying fascination with creation as defined by relation. For Vannier, relation is what lends creation its connective tissue, so to speak, as relation has the ability to cement God and self not only in an outer way, via cosmos and creation, but also in an inner way, via the interior life. Following her conviction that what Augustine is after in his hexaemeron—by which she refers to his practice of interpreting the six days of Genesis mostly in *Confessions*—is not just the external story of creation but his own inner experience, Vannier comes to define the fundamental triad operative in Augustine as *conuersio-creatio-formatio*. This means that she considers conversion to be a precondition for creation, which in turn—as Vannier maintains her stated focus on its inner dimension—leads to formation. For Vannier, the relational aspect of formation comes to full fruition in Augustine's life as a life held together by the arc of conversion.

Augustine's preference for literal exegesis as I have come to define it should not be taken to mean that Vannier's triadic schema unfolds as neatly or sequentially as do the six days. Its threefold course may allow her to gain more insight into the depth of Augustine's soul, but given the endemic restlessness of his soul and the opposing pulls of memory and expectation with which it wrestles, the different steps always become conflated again. The risk of inertia constantly posed by his restlessness puts added pressure on Augustine to keep the hexaemeron's narrative moving along in the clearest and most prudent way. Only loyalty to the letter can accomplish that, for which reason the word-for-word meaning should be the primary reading of the word *literal* in his title *De Genesi ad litteram*, thereby underscoring its narrative dimension.[45] As we see Augustine moving his hexaemeron along, while factoring in Vannier's attunement to relation, it becomes imperative that his hexaemeral interpretation hold together the simultaneity of God's calling creatures into being and his *re-calling* of them to Godself as the goal of their existence. This is all the more important, since without any such re-calling, they would drift off into entropy, forfeiting their being, not unlike in the scenario of Origen's *Peri archon*. Functioning almost as a safety device, only a literal account of creation can prevent creatures from ever vanishing into thin air, its narrative quality both asserting and securing that creation is actually taking place, not merely adumbrated or, conversely, too rashly spiritualized.

To develop a proper understanding of these two points—that God's calling creatures into being is at the same time his re-calling of them, which, in turn, prevents cosmic entropy—it is important to realize that *creatio ex nihilo* is not just a random generative command by which God calls things into material being out of nothingness. In line with Basil's attention to divine will (βουλή), *creatio ex nihilo* represents just as much a *creatio a nihilo*, that is, a gesturing of them away from nothingness. Thus, divine will and its human parallel, the drive for a meaningful existence, thrust creatures forward into the arena of time and history. Augustine states as much in *Gen. litt.* 1.4.9: "Accordingly, where scripture states, God said, Let it be made, we should understand an incorporeal utterance [*dictum incorporeum*] of God in the substance of his co-eternal Word, calling back to himself the imperfection of the creation [*reuocantis ad se inperfectionem creaturae*], so that it should not be formless, but receive a form in accordance with each of the creatures that follow in due order."[46]

For Augustine, in the external scenario of creation's unfolding—my focus on the outward running parallel to Vannier's internal conversion—time and history form the stage where creation becomes ongoing formation, meaning a relatively independent development by which creatures grow and find their place. For while creation comes into being "all at once" (*omnia simul*), as the different individual creatures begin to take possession of their own lives—among which are human beings as "pieces of your creation" (*aliquae portiones creaturae tuae*)—Augustine abides by the letter of scripture in order to indicate and heed the order (narratival rather than hierarchical) of their sequential unfolding.

The letter of scripture is thus the only cosmic order that Augustine is ever prepared to accept, but he does so not just because it is divinely written but, harking back to my earlier analysis, because it is divinely authored. Pushing further the analogy we observed in *Retractations* between the text of Genesis and that of the Augustinian self, I want to clarify that Augustine's respect for Genesis relies in the end more on his view of divine order as a "writerly" one than a written one. The following passage from *Gen. litt.* 1.18.37 elucidates what the focus on scripture's authorship entails. We witness Augustine wrestling with the possibility of a divergence between one's own cause (or view) and that of scripture, as he is faced with a diversity of available scriptural interpretations and an attendant lack of their certainty. Warning us not to insist too rashly on our own view, he counsels: "Perhaps the truth, emerging from a more thorough discussion of the point, may definitely overturn that opinion, and then we will find ourselves overthrown, championing what is not the cause of the divine scriptures but our own [*non pro sententia diuinarum scripturarum, sed pro nostra ita dimicantes*], in such a way that we want it to be that of the scriptures, when we should rather be wanting the cause of the scriptures to be our own."[47]

At first sight Augustine seems to imply here that we should strive hard to appropriate the meaning of scripture as if it were our own. At least, that is the inverse of what he is cautioning against—namely, a forensic imposition of what is clearly our own opinion (*nostra sententia*) onto scripture. But if we understand the analogy between creation and selfhood as revolving around text and authorship, things play out rather differently. For it appears that the deeper issue at stake for Augustine is not any meaning or opinion (*sententia*) but truth itself (*ueritas*). While the message of the divine scriptures (*sententia diuinarum scripturarum*) offers us a shortcut to that truth, scripture's pedagogy of a slow hexaemeral unfolding implies that it is not identical with divine truth, which it nonetheless contains. Scripture does indeed impart truth to the reader, but to the extent that its truth cannot reflect the immediacy of divine vision of *omnia simul*, the reader is unable to access its truth without interpretation.

Temporality, then, is what marks the difference between divine truth, by nature visionary and timeless, and scriptural pedagogy, by nature ordered and sequential. This is clear from God's voice addressing Augustine in *Conf.* 13.29.44:

> To this you replied to me, since you are my God and speak with a loud voice in the inner ear to your servant, and broke through my deafness with the cry: "O man, what my scripture says, say I. Yet scripture speaks in time-conditioned language, and time does not touch my Word [*et tamen illa temporaliter dicit, uerbo autem meo tempus non accedit*], existing with me in an equal eternity. So I see those things which through my Spirit you see, just as I also say the things which through my Spirit you say. Accordingly, while your vision of them is temporally determined, my seeing is not temporal, just as you speak of these things in temporal terms but I do not speak in the successiveness of time (*cum uos temporaliter ea dicatis, non ego temporaliter dico*)."[48]

Surely, scripture's cause or message (*sententia*) is less prone to deception or error than our own, but it also will always need to be unwrapped—its "writerly" intention elaborated—by means of interpretation. What matters is that there is a wedge between the meaning of scripture (*sententia*) and the unadorned truth (*ueritas*) it contains. This leaves the door ajar, however slightly, for communications of that truth other than through scripture. While these communications must, in some manner or other, be aligned with scripture—which, if not always more expressive, is certainly the most articulate rendering given its "writerly" and ordered unfolding—there is no reason to question or disqualify the legitimacy of other voices. Vannier focuses on the self as another voice; I wish to focus on nature's voices. In Vannier's preferred triad of *conuersio-creatio-formatio*, self-examination is the inner avenue that Augustine most fruitfully explores, for which reason she places great weight on coordinating Augustine's hexaemeron with the conversion of self in *Confessions*. Within the scope of my own project it is not only the sequential unfolding of creation as described in Genesis's hexaemeron but the actual fact of its taking place—literally, materially, temporally—that I see as a direct expression of divine order.

Time in or outside Paradise

To give my statements about creation, the temporal self, and the ped-agogy of scripture as continued in literal exegesis more profile, it is useful to dwell on humanity's stay in paradise and how Augustine looks back on this period of ideal yet impermanent bliss for humanity. In her study *Once Out of Nature: Augustine on Time and the Body*, Andrea Nightingale does pre-cisely this, as she lays out how time and history unfold for Augustine. The clock of time, according to Nightingale, begins ticking only after Adam and Eve leave this site of bliss, and the point of history must therefore be to bring them from pretemporal paradise back to its posttemporal equivalent in the guise of transcendent heaven. As a result, Nightingale sees Augustine opting for a starkly disciplined life, which she presents as a devaluation of humanity's embodied life. The aim of the disciplined life, and the goal of the sacrifice it entails, is to prepare humanity for the ideal of transcendent heaven as a substitute for lost paradise. Looking down on earthly life as in-ferior to heavenly existence, Augustine considers leading his fellow Chris-tians down the path of *askesis* as the only way to overcome its limitations. If we reconstruct Nightingale's reading of Augustine on Genesis, whose narrative she extracts without otherwise reflecting on the central mean-ing of scripture for Augustine that it underwrites, God created humanity in paradise as in a standstill place that—because Adam and Eve were im-mune from temporal processes of aging, illness, and death—was marked by eternal bliss. For Nightingale, then, the initial state of humanity was no less than a "transhuman" one, in which they had been destined to stay.[49]

Adam's sin ruined this prospect, and Adam and Eve were thrown into an uncertain future as ordinary human beings. Living temporal lives as humans, rather than remaining outside time as transhumans, they were affected by decay and corruption as their bodies were swept into the pernicious food chain that marks what Nightingale calls "earthly time." Experiencing also a loss of spiritual focus, they concomitantly suffered from *distentio animi*, the distention of the mind or soul that marks "psychic time," which for Nightingale accounts for how they found themselves disrupted mentally. As if killing two birds with one stone, Nightingale sees Augustine's chief concern to be minimizing the disruption of the *distentio animi* or "psychic time" through a regimented discipline of the body in "earthly time." Only with the aid of an ascetic regime that transports us proleptically out of the food chain of "earthly time" by giving us focus on the transhumanism that was our paradisiacal share can we withstand the temptation to sin, even if that temptation can be minimized but never quite eradicated.

For Nightingale, time has no place in paradise but is essentially bound up with the sinful state that defines natural—that is, extraparadisiacal—human life. To prepare themselves for the posttemporal transhuman state of paradise-like heaven, which Nightingale assumes to be Augustine's goal, humans must in their temporal lives suppress not only sexual lust but, it would seem, any passion and emotion, so as to eliminate any disruptions that could exacerbate the damaging pressures of "earthly" and "psychic" time.

As sketched above, Nightingale's is a harsh indictment of Augustine's take on natural life as a constrictive negative spiral of time, body, distended mind, and mortality. For Nightingale, God as acting, as presented in Genesis, may be the deepest source of Augustine's thought, but this makes Augustine no less culpable for subjugating Western civilization to the yoke of this essentially negative Christian paradigm for earthly life. To the degree that her book aspires to free Western civilization from its burdensome Augustinian past, she ultimately aims to do so by deconstructing Augustine's unnatural "nature," insofar as it fails to capture the multidimensionality of human life at the intersection of time and body.

The question is whether Nightingale's diagnosis is correct.[50] A key problem with her view of Augustinian nature is the status of paradise, which she moves outside of time on the grounds that human existence there inhabits transhuman (read: immortal) bliss. But while humanity in paradise does not fall victim to death, it is by no means immune to it. For Augustine, by virtue of having been created mortal, humanity has the capacity to die in paradise even if there would be no need for it to die, while it is furthermore outfitted with various accoutrements associated with temporal existence: a body, need for food, a digestive system, a mind and reflexivity about its use, the capacity to love and bear children, and of course, in a typical Augustinian twist, free will, or the capacity to sin or not to sin. What humans lacked in paradise, then, was not time; rather their stay, in all its temporality, was cut short.[51] For Nightingale, by contrast, the Fall is what makes Adam and Eve from transhuman into human beings by turning their capacity to die into actual death. It also introduces time, insofar as their departure from paradise delimits their lives as held captive by the woes of earthly and psychic time. For Nightingale, in the final analysis, it is death as the benchmark of time rather than time itself that inducts humanity into the food chain of nature, with temporality serving as grist for the mill.

Seeing temporality and death as the twin results of human sinfulness, Nightingale does not only anoint sin effectively as the reason for Adam and Eve's expulsion from paradise, which is a fair reading, but comes to see it also as the reductive essence of their humanity, which is incorrect. Her reading makes temporal human nature as we experience it, that is, cut off from its transhuman paradisiacal past, a strictly postlapsarian affair. Such a complete severing seems misguided to me in much the same way that overly theological readings of Augustine are also (of which Nightingale admittedly is otherwise the opposite), inasmuch as they wrongly claim that it is the Fall, not the taking place of creation, that qualifies humans as humans. This claim goes against Augustine's belief that mortality and temporality underlie both life inside paradise, which God created according to days after all, and life after or outside it. Furthermore, there simply is no warrant for projecting onto Augustine a Platonic distinction between "earthly" and "psychic" time that separates body and soul.[52] What gets conspicuously left out in such a reading is that paradise is also a binding communal ideal for Augustine. This is evidenced in his regard, as early as *De bono coniugali*, of the marital bond between Adam and Eve as constitutive of, and hence continuable in, human society. He deems their paradisiacal community paradigmatic because it is especially close-knit, reflecting not only the partners' essential sociality but also their bond of kinship.[53]

It is tempting to take a bleak view of ordinary life in Augustine, which Nightingale indeed seems to do by painting it in stark contrast to transhuman paradise. Seeing the temporality of natural life as necessarily foreboding death reduces humans to their place in the food chain. The only way to escape this fate seems to be through a kind of reverse overcompensation, as only rigorous ascetic discipline can truly bring one closer to that other transhuman state, heaven or life after death. On this view Augustine's understanding of history simply moves us along from one transhuman state to the next—that is, from pretemporal paradise to posttemporal heaven. This leaves earthly life as an interlude to be merely endured, since it has only inferior, derivative value. That Augustine was the first early Christian author to rehabilitate marriage—preferring it as the social norm to ascetic abstinence for the majority of Christians, and, in doing so, actively aspired to continue Eden's social ideal—goes unnoticed.[54]

From the steely militancy animating normative asceticism, as well as from Nightingale's circular view of history, which allows for no other posture than sublimation in order to escape it, Nightingale's account of earthly life is more Pelagian than Augustinian. In addition her recurring "food chain" simile brings her take on monastic life closer to Origen's model of martyrdom than Augustine's coenobitic friendship.

To refine our discussion of nature, it may be helpful to reflect on two points in particular in Nightingale's analysis. The first is the equation of paradise with a transhuman state outside time, while the second is the idea of natural life as an interlude. As for the first point: while transhumanism can perhaps describe Augustine's view of *heavenly* life, where there is indeed a kind of passing that renders humanity's physical attributes no longer useful, it is not an accurate description of Augustine's paradise. There, humanity lives a carefree life but is no less creaturely for it. Humans are no less temporal or mortal, even if neither time nor death consumes them. What Adam and Eve lack in paradise is the anxiety that makes individual humans unable to synchronize themselves and their actions. This is evident from Augustine's famous outcry, "Late have I loved you" (*sero te amaui*),[55] or his earlier "Give me chastity but not yet."[56] Note that both of these exclamations engage both body and soul, hence nullifying the distinction between earthly and psychic time. Furthermore, Adam and Eve lack the corruption and decay that overshadow extraparadisiacal earthly life, not just physically but, more poignantly, in terms of the grief that they cause.

As for the second point about life as an interlude: it is clear that Nightingale views temporal life, what she invariably calls "earthly" and "psychic" time, as something to be transcended, since she defines the return to transhumanism as Augustine's goal. Yet how do we know that transhumanism is his goal, since it is not clear that it is, in fact, humanity's starting point? Since any determination about paradise and heaven can only be made in this life, it seems that one's valuation of ordinary life is the de facto arbiter of the quality of both prior (paradisiacal) and posterior (heavenly) life. Because it is all we know, and from where we make these judgments, this life must to some extent be normative for how we evaluate the others. In other words, does transhumanism not depend on or derive from humanism/humanity to begin with, rather than vice versa?

If so, and if we focus on created nature alone, leaving the status of heaven aside for now, what paradise is for Augustine is not an atemporal, detached state but a vision of human life lived to the fullest, when time is in step with itself and mortality does not condemn one to succumb to the consumption of death. If we break through the vicious Pelagian circle wherein Christ's redemption must bring us back to paradise or else, humanity's natural life is not crammed between two forms of transhumanism. Instead, in Emersonian fashion, earthly life is the unrolling of one circle and thereafter the starting of a new one. Does it not seem fair to say, then, that paradise is there where nature simply takes place and where that portion of creation that is humanity also has its natural home? If so, then the real problem of temporal human life outside paradise is not merely the threat of death or the food chain but the fact that we are often incapable of feeling at home in our natural lives.

Given that Augustine writes with humanity already having exited paradise, what he must do is to reclaim nature's ability to "take place"—and I return at this point to his practice of hexaemeral reading—which he does by trying to retrace the steps of the divine as it conducted the process of creation. To do so is not a stealth repristinization of the paradise of nostalgia but a way to see ordinary nature for the paradise that it harbors inside itself and, therefore, in the final analysis, also is. This means that Augustine aims to see nature's unfolding as God once saw creation when he said that he saw that everything was good, even very good. As he sets about doing so, Augustine finds that the nature that is there for all to see does not only have a place of its own but also a voice, insofar as the slow, literal reading of Genesis's hexaemeron contains an invitation for nature to speak on its own behalf.

Creation and Conversion 2

My interest in Nightingale's book stems in part from its title. The expression "Once out of nature" is taken from W. B. Yeats's poem "Sailing to Byzantium." Known so well from its opening line "That is no country for old men," the poem invokes a set of tensions that I do not see matching Augustine's.[57] For example, its final stanza—

> Once out of nature I shall never take
> my bodily form from any natural thing
> But such a form as Grecian goldsmiths make
> of hammered gold and gold enameling
> To keep a drowsy Emperor awake . . .

—seems to me more fitting for Origen, who wanted to flee a corporeal life, which he always considered a substitute for spiritual life, by subduing it through a radical ascetic program that drove him to emasculation. In my view, Augustine wants to taste life; thus, he wants to embrace nature rather than suppress it or leave it behind.

Insofar as I have been able to detect, Augustine does not consider actual, physical reality a meager substitute for spiritual creation, nor does his hexaemeron tend in that direction. In fact, Augustine departs from the collective Christian-Platonic tradition, which in Gregory of Nyssa and Ambrose was mediated through allegory and, in a later thinker like Maximus, became galvanized through liturgy. In Maximus, Christology rather than biblical sanction legitimates corporeality, turning nature and the body essentially into sacramental sites, lit up by the dawn of redemption, which is beckoning on the horizon and ready to be celebrated in the ascetic life. Compared to Maximus, Augustine operates on a totally different set of dynamics, as creation's taking place neither expects nor requires such redemption. Nature is there, created by God all at once (*omnia simul*), that is, warts and all, and without seeing material reality as an unintended afterthought, even if its unfolding was serialized in daily installments for us in scripture's temporal pedagogy. Time is endemic to creation, not a side effect of the Fall, though its unruly pressures are. Insofar as sin causes disarray or disorientation, the need to navigate sin-induced pressures—be they temporal, emotional, or social—as best one can is what defines the essential struggle of human life.

How creation unfolds and is ordered in Augustine can be gleaned from his channeling of the hexaemeral scriptural pedagogy in *De Genesi ad litteram*, as well as from the mixture of self-analysis and hexaemeral exegesis he gives us in the final books of *Confessions*. Given the integral link I have posited between nature and selfhood, I see these two modes of reading Genesis—that is, through literal exegesis and self-analysis—as mutually (in)formative. Before returning to Augustine's *De Genesi ad litteram* to focus on the issue of vision and voice, then, let me briefly comment on *Confessions*.

For Vannier, the hexaemeron matters to Augustine because, insofar as she sees creation unfolding through the triad of *conuersio-creatio-formatio*, she regards it as integrally linked to conversion—that is, the self's turning to God, which she regards as the subject of *Confessions*, if not the subtext of all Augustine's works. The role of creation in the process of conversion, for Vannier, is what makes the hexaemeron special, whereas for Nightingale, it is what precipitates Augustine's depreciation of nature, what makes him want to overcome it. To further clarify this role of creation, I want to highlight Vannier's comment that "to express creation Augustine foregrounds anthropology, the relation to the creator."[58] Adding a slight caveat, I think her statement should not tempt us to isolate the self too quickly from natural creation as having a special vantage point, say as *imago Dei*. Rather, in an inversion of perspective, might we not also regard conversion as the self's ongoing attempt to conform to creation's original taking place? After all, as *aliqua portio creaturae tuae*, humanity is an inherent part of God's creation, not a place outside it. This means that for *creatio ex nihilo* to work for Augustine—especially when taken simultaneously in the sense of *creatio a nihilo*, a protective move away from nothingness—it must always keep open the option for self-identity to be not merely autonomous but imprinted on by God, nature, and the unfolding of human material and social life, just as creatures are divinely made. This makes the self's conversion not only a divine beckoning to turn and return to God but also an invitation for the self's further alignment with creation and the creatures that constitute it. Insofar as conversion is not a sealing up but an opening up of a circle, can it not also lead to the linking up of various new circles that tell the stories of creation and its creatures? It matters that the self's conversion is aligned with creation's taking place because it allows us to develop an eye for humanity's role as, on the one hand, a sounding board for the divine and, on the other, a spokesperson for creation as it reaches out to the divine.

The latter point implies that conversion is a much more comprehensive process than the self's return to God envisioned in such a way as to use creation *merely as a means* to that end. This latter envisioning is the natural result of a reading of Augustine that depreciates creation necessarily, such as Nightingale's. In line with my insistence on creation's taking place, on God's carving out a place for it that we cannot only call natural but is fully its own, humanity signals through conversion that it takes seriously the need to synchronize itself with divine demand, which presupposes that humanity accepts itself as created.

While the latter observation seems unsurprising, its corollary in nature has remained underemphasized—namely, that there is the possibility that creation can communicate on its own without the screen of humanity acting as its interpreter. In parallel fashion to *Confessions*, then, I would argue that

De Genesi ad litteram may also be called a confession. That is, it is inherent in God's *creatura* to allow it to be the site of *confessio Dei*, as a kind of turning and expression of the world in which humanity is included. Thus, what Nightingale's motto, "Once out of nature," lifted out of the wider meaning of Yeats's complex poem, fails to capture about Augustine's creation is the following: first, that created life is literally a full life, since through it, God gifts humanity time as a platform for an array of human passions and emotions including celebration and joy; and, second, that Augustine thereby opens the possibility that nature, whose taking place is narrated in hexaemeral form, undergoes its own conversion, meaning that it is not just a story to be told but that it has its own story to tell.

Vision and Voice

Vision in Augustine, as I claim in this final section, has from the beginning a deeply eschatological pull to it. When God sees that all is good at the very beginning of things, this vision harbors the goodness inherent in creation itself and in doing so anticipates the sabbath's rest, when such goodness will no longer be disturbed. Vision is thus, in part, eschatological. Voice, however, is not, located as it firmly is at the beginning, when God speaks creation.

Vision is the dominant theme in book 12, the concluding book of Augustine's *De Genesi ad litteram*.[59] Taking a step back, Augustine clarifies to his readers that in this book he departs from the word-for-word exegesis of the earlier books, as he delves more deeply into the question of paradise's location. He does so specifically by relating it to the scriptural fact that Paul knew a man in Christ who was snatched up to the third heaven (2 Cor 12:2) yet did not apparently know whether this happened to him in or out of the body. This leads Augustine, via the question of how Paul saw paradise, to explore the different kinds of vision, layering them by means of his exegesis of the phrase "you shall love your neighbor as yourself" (Mk 12:31) on three different hierarchical levels. There is, first of all, vision with the eyes as we read the letters on the page, which Augustine calls *bodily vision*. Next, there is vision with the human spirit when one imagines the neighbor mentioned in the text, which he consequently calls *spiritual vision*. Third, there is vision that requires the attention of the mind, as when in reading the phrase at hand one comes to understand love itself, which he indicates with the term *intellectual vision*.

While Augustine qualifies his statements about this division at the end, telling us that others may call "soulish" what he calls spiritual, and spiritual what he calls intellectual,[60] it seems that intellectual vision may well be what Paul's witness experienced, as his vision seems to have been observed in the mind alone, entirely apart from the senses. But is the mind therefore out of body and, for that matter, out of nature in its materiality?

In clarifying what is seen there, and with an eye toward God's comments at the beginning that he saw creation as very good, it appears to me that Augustine doubles down on the fact that paradise *is* reality, instead of being driven by a transhuman need to move away from it. Creation is a stepping stone, allowing us access to the creative Word itself by forging a direct unity with it, if we could only move beyond the bustling sounds of voice to a contracted stillness, as in the famous vision at Ostia from *Conf.* 9. Augustine then lays out his understanding of the third heaven in terms of intellectual vision: "We consider then that that is where the apostle was not unsuitably snatched away to, and where the paradise is, that is better than all others, the paradise of paradises, if we may so call it. If the good soul, after all, finds joy in the good things in the whole of creation, what could outdo the joy which is to be found in the Word of God, through whom all things were made?"[61]

Ecstatic vision thus *affirms* the status of creation rather than bypasses it. Making a transition back now from the ethereal heights of this ecstatic vision to Genesis's profile of newly minted creation itself, it seems to me that the claim of Augustine's hexaemeron is not that God created paradise out of nothing as a standstill place for transhumans to dwell in eternal bliss. Rather, how I read Augustine's *De Genesi ad litteram* is that created paradise marks God's fundamental decision to henceforth manifest Godself through creation as a stand-alone product, a world unto itself. This makes the work from the start a deeply dialogical affair, since creation's relative independence, just like humanity's, makes it yearn to engage in conversation. It is dialogical in a different way than is *Confessions*, since in *De Genesi ad litteram* the creative Word is not searching for a liturgical or other kind of antiphonal response, even if in other ways this work is also a kind of confession, but is satisfied with the acknowledgment that it is performed by tangible creation.

One way in which this modified material dialogical aspect appears to have shaped the work is that, its length notwithstanding, Augustine has more directionality in unfolding the story of creation here than do some of his Greek counterparts. In contradistinction to the Christian-Platonic tradition, including Maximus but also the later Eriugena, there is not the same kind of leisurely *processio* that only reluctantly, under pressure of sin and the Fall, turns into the *reditus* that must follow. This implied urgency, tied to a heightened sense of temporality, may be another reason why spiritual creation does not precede corporeal reality in Augustine; rather, all of creation comes into existence at once, even as we must not forget that in his concluding excursion intellectual vision appears to underlie and sustain it all along. The urgency of this task means that corporeality simply requires Augustine's deepest attention, needing to be assessed from every side as he breaks down and unpacks the notion of *omnia simul*. At the same time— and this is the other side of the dialogical equation—it is important to realize that it is only creation's genesis in response to the call of the divine that draws God out of his solitary confinement and into relation.

We find an eerie echo of divine loneliness in God's command "Let there be light," leading Augustine to speculate: "And who was there, who needed to hear and understand, to whom this sort of utterance would be addressed [*Et quis erat, quem oportebat audire atque intellegere, ad quem uox huius modi proferretur*]? Or is this an altogether absurd and literal-minded fleshly train of thought and conjecture?"[62] It is surely fleshly conjecture that Augustine gives us here, the sort that would not have occurred so easily to Origen or Ambrose. But in the context of Augustine as an author writing a book for a communal readership, it is anything but absurd. The question makes the voice of the divine, which speaks creation into being, simultaneously reach far beyond. It is as if God steps out of the hexaemeral frame for a moment to touch base with his interpreter Augustine and his readers directly.

From one angle, then, it appears that this passage reasserts the parallelism in *Retractations* between two different kinds of texts, the writing of self and scripture's hexaemeral writing, both needing to be read and decoded by an astute readership. On a deeper dialogical level, the dynamics have subtly but radically shifted, as the juxtaposition of texts signals that through actively speaking, God is reaching out to creation, freeing it to perform the divine Word and thereby allowing it to take place in relative independence. The two partners, namely God and creation—and this is the essence of what I see as their dialogue—remain all the while in conversation, that is, inseparably connected. Thus, what Augustine seems to want to convey through his hexaemeral commentary is, first, that there is no creation without God's speaking—no light, no heaven, no water, no animals, no human beings—but, second, that the divine Word is ineffective when creation is not being continuously and materially performed—that is, taking on a life of growth and development away from the divine. Even if creation from nothing, in the sense of surreptitious emergence, is not what primarily matters, the notion of *omnia simul* signals that it does matter by securing that all things and levels do indeed materialize out of nothing, not having any previous levels or realities to emerge from since they all emerge at once. But the heart of created existence is its concrete, material, and continuous performance of the divine Word. A greater threat for creation *qua* creation than the disappearance back into nothingness is the disappearance into the overwhelming light of divine vision where all things are very good. It is creation's performance alone—material, repetitive, continuous—that allows it to maintain its separateness by making sure that it stays "in place," there for all to see without any threat of vanishing back into the thin air of divine Spirit. With its patent voice and patient pace only literal exegesis has the dispositive quality to bring this separateness out.

Divine speaking, then, especially the Word of creation, does not only bind God to Augustine or the human self, either of whom can be the addressee in the above passage, but on a deeper dialogical level it binds God to all of creation. For creation becomes invited to perform the divine words, so to speak, whose beauty finds reflection in the concrete arrangement of individual creatures, as hexaemeral text becomes instantiated in and continues to weave what is creation's own story. It is that concrete arrangement, that textuality of creation being woven and the hexaemeron empowering it to keep weaving, that allows creation to speak for itself. Creation speaks not so much by emitting its own words as by bearing them out, projecting signification of the kind that need not only be passively receptive of the divine Word but is understood and encouraged to be active, even proactive.

Insofar as creatures can connect with other creatures in coalitions and alliances of their own, they extend and give rise to various forms of community, as temporal nature bends to the arc of history. While Augustine's nature is not a liturgical universe, his is certainly a polyphonic world, one in which, held together by the hexaemeron as the melody of a Bach chorale, creation is free to perform and modulate its own eloquence.

Nature as Conversation

Augustine on Verbal and Natural Signification

To take stock of where we are at the end of our first foray into that other, more elusive nature, I want to elaborate in this postscript on my idea of putting nature into dialogue or conversation. While I have done so through debating an array of premodern authors on their view of nature and creation, the focus inside my chapters has increasingly been on nature's dialogue with God and with humanity. Such a dialogue occurs in Eriugena's *Periphyseon* ever more poignantly, even to the point, I contend, of transforming nature itself into conversation. To explain nature's transformation from being in conversation to being itself conceived as conversation, we need to discuss Augustine's complex view of signification, which is visual—as we ended the last chapter with modes of vision—but also vocal-verbal, which opens up the connection with conversation. Since vision in Augustine is mostly eschatological, I will pursue here the vocal-verbal aspects of nature. From there I will move to Eriugena and the idea of nature as conversation.

While the study of nature in premodern thinkers is often found linked with the divine through Paul's statement in Romans 1:20 that God's invisible nature is attested through his visible attributes,[1] making the main prism through which to see nature that of divine self-disclosure, various references in *Confessions* can make us more attuned to the idea of nature's own voice. Below I cite two examples from *Confessions* where nature is associated with speech.

In the first example, from *Conf.* 9, we are privy to Augustine and Monica's last dialogue before her death. In the so-called mystical vision at Ostia, we follow their gradual move upward from earthly, material reality to the ethereal heavenly reality of the divine. The passage in question does not present created things as actually speaking, for the general emphasis is on vision above voice. Insofar as Augustine imagines hearing them speak, however, he tries to infer what it is they would say: "We did not make ourselves, we were made by him who abides for eternity." This silence and the creatures' general lack of language enables Augustine and Monica to lock nature out of their shared bond of human intimacy. Thus, we are confronted with the limits of nature's relationality, as the interlocutors in this passage seem keen on bypassing creatures to hear the voice of the creator unfiltered and unmediated.

This is how Augustine records their vision as he converses with Monica:

> Therefore we said: If to anyone the tumult of the flesh has fallen silent, if the images of earth, and water, and air are quiescent, if the heavens themselves are shut out and the very soul itself is making no sound and is surpassing itself by no longer thinking about itself, if all dreams and visions in the imagination are excluded, if all language and every sign and everything transitory is silent—for if anyone could hear them, this is what all would be saying "We did not make ourselves, we were made by him who abides for eternity" (Ps. 79:3, 5)—if after this declaration they were there to keep silence, having directed our ears to him that made them, then he alone would speak not through them but through himself. We would hear his word, not through the tongue of the flesh, nor through the voice of an angel, nor through the sound of thunder, nor through the obscurity of a symbolic utterance. Him who in these things we love we would hear in person without their mediation.[2]

From the fact that creatures are seen as unable to speak, we can nevertheless infer that they equally desire to have a voice of their own and to signify who they are through self-expression. Augustine makes a similar point about creatures' inchoate attempts at linguistic self-expression in the related passage below from *Conf.* 10, where he asks what he loves when he loves his God. Here he engages in actual dialogue with natural creatures, though, in describing the answer that they resonate back to him, namely their beauty, he mixes registers again by slipping in the visual with the vocal:

> And what is the object of my love? I asked the sea, the deeps, the living creatures that creep, and they responded: "We are not your God, look beyond us." I asked the breezes which blow, and the entire air with its inhabitants said: "Anaximenes was mistaken; I am not God." I asked heaven, sun, moon and stars; they said: "Nor are we the God whom you seek." And I said to all these things in my external environment: "Tell me of my God

who you are not, tell me something about him." And with a great voice they cried out: "He made us" (Ps. 99:3). My question was the attention I gave to them, and their response was their beauty [outward form (*species*)].[3]

While it is tempting to dismiss this passage as a pagan rudiment indicative of an older material notion of God, Robert Markus has productively connected the visual and the vocal by highlighting the Augustinian link between signification and creation. Exploring that link, this postscript will press on the notion of verbal or vocal signification to see what room there is in Augustine and Eriugena for the notion of dialogue and conversation with nature.

Markus's *Signs and Meanings: World and Text in Ancient Christianity*, a collection of several essays on Augustine's sign theory,[4] contains perhaps the most complete analysis of Augustinian signification. These essays are held together by Markus's deep interest in the multivalence of scripture or what he calls, using an expression from James O'Donnell, "the polysemy of scripture and the licit plurality of interpretation."[5] For Markus, Augustine's openness to plural interpretation is rooted in the new network of meanings that he culls from the foundational distinction between *res* (things) and *signa* (signs). Through this distinction, first introduced in *On Christian Teaching*, Augustine multiplies the possibilities of scriptural exegesis exponentially. For, pushing the notion of "sign" beyond the literary sphere, he stretches its exegetical application into the natural sphere. Expressing his awareness of this, Markus holds that when things (*res*) are seen as signs (*signa*) in Augustine, they do not automatically become swept up in the realm of verbal interpretation but retain their materiality rather than forfeiting it. The fact that even in their capacity as signs, material things are not deprived of their thingness allows Augustine to engage nature in what is in fact a double conversation. By this I mean that, parallel to a vertical interpretation along the conventional lines of things and their spiritual or conceptual meanings, there is a new web of interpretations that allows for the horizontal interconnection of material things themselves.

As Markus clarifies, by giving signs a hermeneutical purchase that goes beyond the literary, Augustine coins in fact a novel usage of signs and therefore of scriptural exegesis.[6] Whereas things (*res*) can indeed function as literary signs (*signa*) through which to communicate spiritual or intellectual matters, as physical signs retaining their thingness they can, through bodily and material communication, also point to other material things. Instead of seeing things *qua* signs and their (literal or spiritual) meaning as inhabiting two separate, vertically ordered spheres, Augustine's referential system unlocks a far richer network of meanings with myriad interconnections. Especially, Augustine allows for things *qua* signs to not only be verbally explained but also to gain further clarity through correspondence with other things.

In contrast to the upward pull of Origen's Alexandrian tradition, Augustine bequeaths us a solid preference for the literal sense and the material and historical groundedness of things that underlies it. Pointing us *forward* as well as upward, Augustine's hermeneutics of signification is suffused with incarnation and shot through with temporality. Given the dialectic of promise and fulfillment animating both (incarnation and temporality), it bears little surprise that Markus's particular focus, which parallels Augustine's own, is on deeds or events that draw together the Old and the New Testament. As he explains: "A text can be taken in its strictest literal meaning and may nevertheless have, indirectly, a further reference. It is the events narrated by it that may themselves have a meaning, that is to say, they constitute, as he [Augustine] said, a divine discourse. It is not the biblical text that means something other than what it appears on the face of it to be saying. It is what the text is literally referring to that itself has a further meaning."[7] Below I extend Augustine's notion of a divine discourse (*diuina eloquentia*) to make it resonate not only with the deeds or events on which Markus focused—and what Augustine in *De ciuitate Dei* 11.18 calls the eloquence of things (*eloquentia rerum*)[8]—but also the world of natural things.

If we inspect what biblical interpretation looks like for Augustine with regard to the so-called *signa translata*, the figurative or transferred signs that are hardest to read and thus most in need of interpretation, we see Augustine lay out consecutive stages of meaning. The word *ox*, to take Markus's example, can indicate the physical beast (scenario A), but once the sign is known, "ox" (that is, the word and/or the beast) can also refer to the evangelist Luke (scenario B).[9] For a thing to be a sign, there has to be a first meaning, but by reverberating onward, the thing-that-is-a-sign can itself also acquire a second meaning. That second meaning may be called transferred (*translata*), but it is still embodied and temporal. Markus visualizes the train of interpretation that emerges as follows:

A: Sign (word) → Signified (thing, event)

B: Sign (word) → Signified_1 (thing, event)$_1$ → Signified_2
   [Transferred sign = Sign + signified_1] → Signified_2

where A represents the normal relationships of signification
and B represents the relationships in the case of 'transferred' signs.[10]

Augustine's notion of a transferred sign compresses what, on unpacking, unfolds into a double meaning, corresponding to his engagement of creation in what is hence a multimedial conversation—that is, through words and through creatures. Since the process is set in motion by one and the same sign—the word *ox* can denote both the animal and the human evangelist—the assumed priority of anthropological interpretation retreats into the fuller resonance of creation.

In the exegetical context of his stated focus on "the polysemy of scripture and the licit plurality of interpretation," Markus applies Augustine's notion of *signa translata* mostly to signs as words. But there is no reason why the chain of resonances should end there, as material or created things (*res*) can themselves evoke further meanings. In the absence of a hard and fast distinction between verbal and nonverbal signs, Augustine's hermeneutics of signification allows him to cast a wide net of meaning over all of created reality, even though the net of natural(ist) interpretations remains replete with scriptural echoes.

Although the ostensible purpose of Augustine's *On Christian Teaching* is to teach scripture, reading the work in Emersonian fashion, with an eye toward the wider hermeneutical circles that it draws, helps us see that Augustine is unlocking biblical exegesis here as a supple and capacious genre of interpretation of the broader natural world as well. I begin to diverge from Markus in that I see Augustine, apparent in his many Genesis commentaries, as being interested not just in "the polysemy of scripture" as a theoretical possibility but in actively upholding a full spectrum of dynamic meanings of biblical and natural signs *qua* signs. I submit that the full spectrum of meanings is more interesting to him than finding the single most accurate interpretation, this being the interest of Jerome and his peers. It would be a fallacy to conclude, however, that Augustine leaves interpretation up to the whim of the interpreter. As Markus explains, Augustine's hermeneutics of signification as the process of exegetical meaning-making is always patterned on a clear triad, consisting of (a) the signifier, generally seen as the created thing, (b) the signified, that is, the message that it sends about divine authorship, and (c) to whom something is signified, that is, the reader(s) or audience. There is no reason to think that, just as in Maximus's parallelism of nature and scripture, a similar triad cannot apply to the reading of natural signs.

Augustine's hermeneutical system is so powerful because it makes sure that signs (*signa*), by which Markus refers especially to the *facta* or events of the Old Testament but in which I have now included natural things, retain their maximal expressiveness. As Markus sees it, Augustine's arguments bring him at times into conflict with the Jews of his interpretation, whom he considers bound to useless signs in carnal fashion, that is, heeding the commandments without adequate motivation or reflection, and at other times with the pagans of his interpretation, who confront him with the insidious problem of idolatry—that is, the worship of *simulacra*. Because, in Augustine's view, pagans lack any recognition of God as the author of creation, they mistake creatures and their artifacts for the creator. His awareness of idolatry leads Augustine to state in *Conf.* 10.6.9 that the mission of created things is to reflect their maker, and they can only do so in relation to a reader who has the right semiotic intention. Hence, while Augustine's hermeneutics blurs the divide between scripture and reality (by which we mean culture and nature), the interpretive circles he draws remain concentric in reflecting a joint sense of purpose. Fostering creation's worship of God, through his theory of signs, must blend in with the *Confessions'* overall goal of conversion, in the course of which the world and the self are slowly but surely reoriented to God. Much like Augustinian selfhood, creation (meaning the panoply of creatures) in *Confessions* does not merely intone a state of divine dependency but actively participates in the work's ongoing process of interlocking dialogue, prayer, confession, and the dynamism of conversion.

While Augustine never states that created things are signs of, and hence "speak," their creator, the triadic theory of signification put forth in *Conf.* 10.6.10 that I see operative in the *Confessions'* last three books (11–13) allows us to weave an ever-closer relational web between the three parts of the triad: the reader/interrogator (to whom is signified; we might say the confessing self), the creatures put to the question "who made you?" (the signifiers), and the creator (the signified). Within this triadic relationship, the statement that creatures *qua* creatures speak of their maker does not so much connote a doctrinal or exegetical truth as it instigates a performative one, insofar as creatures are called upon to execute the so-called *eloquentia rerum*.

For Markus, the triadic configuration of subject-sign-signified is an important mainstay of Augustine's hermeneutical thought. It reflects his habit of reading not just referentially (that is, accurately denoting any signified objects) but ethically (that is, connoting that signification thrives under the right relations). If we take account of the full spectrum of Augustine's hermeneutics, its depth alongside its integrative reach and moral character, and bring it full circle, it reflects the circle of signs proceeding out from God and returning back to God. Given this parallel, it becomes clear that there are great repercussions for the interpretation of the natural world. Augustine alerts us to the need to communicate creation's relevance and meaning, simultaneously charging us with a sense of both duty and possibility. As Markus states:

> The ability to read God's deeds in the Old Testament as his speech (*eloquentia diuina*), and the ability to read God's creatures in the world as telling of their creator (*eloquentia rerum*), both require the right disposition or intention as Augustine would say. The disposition to see things as signs, with a meaning beyond their obvious, immediate appearances is an intention to "put them to the question." Meaning is not obvious to us: our understanding is clouded. Fallen human beings as we are, we are permanently liable to failing to communicate and failing to be communicated with. In this life we are denied the transparency of mutual understanding which would allow direct communication between us and other minds.[11]

I am not offering a correction to Markus but am further applying what he says. He limits things (*res*) mostly to deeds and Old Testament events; I run with this and apply it to all creatures. Given that humanity's understanding is clouded, what we search for is to read the opaque sign as one that discloses meaning and therein enables and inspires us as flawed human beings to break through life's opacity by conversing with others.[12] The difficulty implied in the search for meaning makes clear how much Augustine's quest is at root a quest for transcendence: transcendence not of creaturehood in Neoplatonic fashion but of the reification of compromised selfhood. His wide exegetical circle invites the sinful self to open itself up to a fuller manifestation of the divine. Finding itself imprisoned among the opaque signs tends to isolate the self from its human community as well as from the realm of creatures, to which it also belongs and craves access. Right interpretation, in other words, whether scriptural or natural, means conversion to community—envisioned as a fully embodied, nature-wide, incarnational reality—in which humans have their place with all of creation: a community, consequently, in which nature is not only freely thought but in which it can freshly raise its voice and speak its mind.

Nature's Conversation in Eriugena

If we transfer the idea of natural signification to the *Periphyseon*, are we not justified in seeing Eriugena's monumental work as an attempt to encode the eloquence of things (*eloquentia rerum*) or, better, nature(s)? As for Augustine, the extraction of meaning from natures that speak of their creator marks also for Eriugena an important step toward a more complete understanding of God's eloquence (*eloquentia diuina*), which Eriugena likewise sees conveyed by scripture. In line with this I submit that the *Periphyseon* is best read as an early medieval application of Augustine's referential system of signs and things, with renewed emphasis placed on divine communication directly through *res* (things)—that is, natures- or creatures-as-signs—as much as through verbal signs. Pursuing this approach will be particularly helpful in connecting the *Periphyseon*'s first half, consisting in a dialectical appraisal of reality that includes a wide-ranging discussion of the applicability of the Aristotelian categories to God, of denial and affirmation as the two modes of Dionysian theology, and of theophany, with the work's second, exegetical half, in which a literal interpretation of the hexaemeron is followed by an allegorical one.

Approaching the *Periphyseon* from the perspective of signification and communication, we do best to see the dialogue partners, *Nutritor* and *Alumnus*, as stand-ins for the community at large, meaning the integrated whole of human and other created beings, to whose restoration both Augustine and Eriugena aspire. Teasing out further implications, I want to underline that community in Eriugena includes all of creation, as I think it also does in Augustine. Creation or nature as community mirrors the comprehensiveness of Maximus's liturgical cosmos, even if it remains true for Eriugena and Augustine, that knowledge is primarily communicated through the direct interaction of human minds with each other. Maximus's overt liturgical agenda and requisite ascetic setting, however, are absent in both Augustine and Eriugena. Even so, exegesis lies at the heart of what all three of them do, and salvific intent remains their key operative principle.

So, what kind of restoration does Eriugena have in view? The liturgical ambiance that dominates Maximian discourse is drained from the

Periphyseon, while the Carolingian school context lends the dialogue of master and student a formulaic playfulness.[13] These are connected in that we see the role of Christology, in Eriugena, becoming transposed to anthropology (rather than liturgy) in a manner proportionate to the work's school-based, protohumanistic context. Following Gregory of Nyssa's idea that humans are made in the image of God, Eriugena sees the human mind (*nous*) as the site of humanity's quality of *imago Dei*, even as he complements it with the Maximian trope that humanity is the workshop (ἐργαστήριον/*officina omnium*) of creation.[14]

Debating the possibilities and limitation of their knowledge, master and student enunciate at one point what is no doubt Eriugena's most radical position on the unity and priority of divine understanding—namely, that man is a notion eternally made in the divine Word (*homo. . . . notio aeternaliter facta in mente diuina*).[15] While at first blush this statement underwrites Eriugena's so-called idealism, insofar as it privileges the essential notions in the divine mind over the corporeal existence of things in the world (and hence seems to disqualify human efforts to arrive at rational judgment through sense-based knowledge), the *Periphyseon*'s latter half, the half that deals with the return through exegesis rather than speculative analysis, shows the situation to be rather more complex. Not unlike Emerson, Eriugena stays remarkably close to the concrete and the imperfect, which means that he often brings into focus the sinful and imperfect aspects of human life. Besides seeing humanity as a notion inside the divine mind, master and student also put forth another definition of humanity, according to which the human mind contains within itself the images of all the things that it knows.[16] In the same Platonic vein, humanity's understanding of things must rank above their historical, material reality.

These two definitions clash on the point of human self-knowledge: is it perfect in its identity with God's transparent knowledge, or is it deficient in its rootedness in the imperfect process of human knowing? Eriugena struggles to achieve a resolution. Making ingenious use of Dionysian *apophasis*, he grafts what amounts to the divine epistemological surplus (by which humanity is a notion in the divine mind) onto humanity's epistemological deficit (whereby its imperfect state impedes perfect self-knowledge), canceling out their opposition. Yet in the quest for knowledge, humanity's imperfection ultimately has the edge, insofar as the dialogue between the two interlocutors produces, and hence relies on, intersubjective knowledge.

With intersubjective knowledge taking priority over both imperfect
human self-knowledge and angelic intellectual transparency, Eriugena secures that the interaction of master and student can yield concrete mutual
understanding. In this way Eriugena turns the ontological asymmetry of
human and divine into what is a level epistemological exchange between
two human interlocutors. This gives the unfolding of *nature* the status of
template in ways reminiscent of how Augustine saw *scripture* as mediating
between God and humanity.[17] In this epistemological exchange the perfect
divine creator and the flawed human self can through intersubjective dialogue likewise interact, thereby contributing to nature's unfolding. Thus,
every step forward in the dialogue allows Eriugena to express nature's role
with greater accuracy. Insofar as master and student jointly move the course
of *natura* forward, their dialogue sustains and even expands *natura*'s orbit by
instilling in us the sense that, short of the eschatological return, the point
of nature, and its entire span, is conversation.

The broad restoration of community, including the full community of
creation, that the *Periphyseon*'s dialogue aims to achieve liberates nature to
both speak freely and be freely thought. Nature can present itself *qua* nature
to the two discussants and, in and through their intersubjective exchange,
to both God and humanity alike. This does not mean that nature ceases to
be the go-between that links God integrally to the human self, for it will
continue to fulfill that ambassadorial role; but in Eriugena, more than in
any other premodern Christian author, *natura* as a genus is itself permitted to take the initiative. It can style itself radically, even if it always also
remains elusive, presenting itself on its own intractable terms.

I clarified in Chapter 1 that God is internal to Eriugena's *natura*, as
genus that encompasses all things that are and are not, but over the course
of the work's development we see *natura* becoming saturated with both divine and human aspects. In consequence, not only do we see nature lead
the way for both, but it begins to shape the manner in which they relate
to each other. It is not simply the case that God's inscrutability migrates
from the divine to the world of nature, for inscrutability remains a prime
divine attribute. Rather, it is as if inscrutability becomes extended to the
whole of nature, which, since it encompasses the divine, can absorb it as
an endowed trait. The goal of master and student is in my view not solely
to find unity with each other and God, as I have earlier rejected the mystical alongside the idealist option, but the establishment of coexistence and

open communication between God and humanity and, in equal measure, between *natura* and the natures that constitute it. In the final analysis, then, "thinking nature" means that *natura* is accorded the space to define itself. Nature's self-definition, in turn, affords us, through our following of the interlocutors' effective exchange, both the opportunity and the platform to (re)constitute ourselves and our relationship to God by adding the natural to the scriptural world.

Defining the purpose of Eriugena's *Periphyseon* as the gearing up for such an all-out and holistic communal restoration—not only of master and student with each other but of both with *natura* as including the divine and the physical world—helps us to better situate *natura* in both its temporal specificity and its projection of omneity. While the Augustinian conversation of individual creatures with the divine by means of the discourse of things (*eloquentia rerum*) takes us outside of scripture, what the concept of *natura* does for Eriugena is precisely to keep us inside of its all-encompassing contours as conveyed by the *Periphyseon*'s wide-ranging and at times meandering text. Mediated by the interlocutors' reciprocal understanding, *natura* bolsters the long-sought contact of creation with the divine, not through mystical ecstasy but by doubling down on the well-worn path of dialogical conversation: between master and student, between creator and creatures, between God and the human self. Return must hence always be aspirational. It never comes to rest, not even in God, who is not created and does not create. It is dynamic continuity without end.

The Vision from Nature's Conversation

The idea the *Periphyseon* projects—that when thought freely, *natura* unfolds as aspirational conversation—has far-reaching consequences. To the extent that Eriugena's *natura* is rooted in dialogue, it is not just a platform for conversation; it shapes and becomes itself that conversation. Different from Augustine's known penchant for soliloquy, even if likewise dialogical in character, *natura*'s conversation in Eriugena is fluid and externalized, at once ongoing and ever deepening.

That *natura* is all about and, in the end, indeed, *is* conversation is most clearly expressed in those moments when we see the need for a course correction, as when the *Periphyseon*'s power structure suddenly shifts or its intuitive balance becomes endangered. At such pivotal moments, where there may be the possibility that nature's course becomes derailed, Eriugena must use all his rhetorical mettle to try and keep the *Periphyseon*'s conversation, and hence the course of nature, on track. I conclude this postscript with a few salient examples from the *Periphyseon* of the intertwined fate of nature and self; I will focus especially on the suspense that is generated when God's favor is on the line.

My first example is when God confronts Adam in the garden with the haunting reproach: Adam, where are you (*Adam, ubi est*)? This moment reflects an acute crisis not only for human selfhood but also for creation, since doubt is cast on Adam's integrity as made in God's image and, as such, the cornerstone of creation. Eriugena's reading of the biblical episode as one "fraught with background" to use Auerbachian terminology,[18] is truly magisterial. When an intervention in the discourse of things (*eloquentia rerum*) is urgently needed to avert escalation, Eriugena avails himself directly of divine speech (*diuina eloquentia*) by means of an exegetical voice-over, which we can see as a move that collapses a commentary on the scriptural text, with God stepping out of the scriptural frame and into that of the *Periphyseon*. Even as God's pressing accusations intensify to register outright conflict, the unfolding of *natura* continues unflinchingly:

> In this intelligible Paradise God goes walking. For He is the Guardian and Inspector of the Garden which He has made in His image and likeness. His is the Voice which cannot be expounded: "Adam where are you?" This is the voice of the Creator rebuking human nature. It is as if He said: Where are you now after your transgression? For I do not find you there where I know that I created you, nor in that dignity in which I made you in My image and likeness [*non in ea dignitate, qua ad imaginem et similitudinem meam te feci, inuenio*], but I rebuke you as a deserter from blessedness, a fugitive from the true light, hiding yourself in the secret places of your bad conscience, and I enquire into the cause of your disobedience [*causamque tuae inoboedientiae inquiro*]. Do you suppose that I do not know what you have done or whither you have fled or how, in fear of My voice, you have concealed yourself or in what way you came to a late recognition of your nudity, that is, of the purity and simplicity of nature, in which you were created? Have you not gone through all this because you have eaten of the tree which I commanded you were not to eat? For if you had not eaten perhaps you would not fear the voice of your Creator as He walks within you [*non fortassis uocem deambulantis in te creatoris tui timeres*], nor flee from His face, nor have become aware of the nakedness which you lost when you sinned.[19]

Walking a thin line here between divine accusation and his own vicarious indignation, Eriugena faults Adam, even when Adam places the blame for sin on Eve. Whereas Augustine held Adam and Eve jointly responsible for human sin,[20] Eriugena declines to settle the matter within their marital union alone, relating everything back to the overarching story of *natura*. Yet again, it becomes apparent that for Eriugena, the integrity of nature's unfolding as the eloquence of things (*eloquentia rerum*) is what counts above all. When both human and divine eloquence are absorbed in the story of *natura*, its flow of procession and return cannot be disturbed.

Refusing to give Adam a way out, Eriugena applies another exegetical voice-over, this time folding what seems like an outright derailment of nature into its ongoing procession, making it just another one of nature's many folds and creases. Since God obviously saw through Adam's deception in blaming everything on Eve, Eriugena looks back on Eve's creation ("Let us make him a helpmate") as, in retrospect, a case of supreme divine irony:

> This is also made quite clear by God's ironical words [*diuina ironia*]: "It is not good for man to be alone. Let Us make for him a companion like unto him" (Gen. 2:18). The meaning is: Man whom We have made in Our image and likeness does not think it good to be alone, that is, to be a simple and perfect nature, abiding everywhere without the division of his nature into sexes, being wholly in the likeness of the angelic nature [*uniuersaliterque diuisione naturae in sexus, ad similitudinem angelicae naturae, absolutum permanere*], but prefers to tumble down headlong into earthly couplings like the beasts and so to multiply out of his seed the unity of his nature through carnal generation and the sexual organs of his body, holding in contempt the mode of propagation of the heavenly host. Let Us then make for him a companion like unto him through whom he can perform what he longs to do, that is to say, a woman who is fragile and unstable like the male, and is eager for earthly lusts [*feminam uidelicet, quae similiter ut masculus fragilis ac lubrica terrenas appetat concupiscentias*].[21]

At no point does Eriugena deem Eve's creation inferior or in any way less real, or less willed, there being no trace of idealism, lofty Platonism, or male ascetic dominance. For Eriugena, God's irony marks the acknowledgment of physical reality rather than a detraction from it. Closer to Augustine than to Gregory of Nyssa, Eriugena gives us a wholesale embrace of humanity's predicament.

Yet precisely when *natura* is at an ever-greater remove from God, Eri-ugena reins it back in. The point I want to bring across is that his exegeti-cal choices carry real weight for the unfolding of *natura*, forcing him at times to interpret scripture against the grain so as to accommodate *natu-ra*'s progression. Nature continues its autonomous course even when the possibility for return appears blocked. Adam's expulsion from paradise is a case in point, as Eriugena interprets this consummate hurdle for return as a de facto promise of humanity's ultimate restoration. His crafty pivot from protology to eschatology has been recognized as enabling humanity's turn from procession to return.[22] In changing the meaning of condemna-tion (namely, that man will not stretch out his hand to eat from the Tree of Life) to promise (that one day he may eat from it), Eriugena sets the stage for the promise of paradisiacal union of nature and self:

> The Return of which we speak is implied in the Voice of God saying [*Di-uina itaque uox talem reditum, de quo sermo est, his uerbis insinuat*]: "Now therefore (*Nunc ergo*)," or as the alternative translation more explicitly puts it, "And now (*et nunc*), said God." . . . Do you see the largeness of the Di-vine Compassion which is compressed within the single temporal adverb Now and a single causal conjunction Therefore? This same divine compas-sion, converting the lamentation for man to a consolation both of man himself and of the Heavenly Powers, promises under an ambiguous and interrogative form of speech the Return of man into Paradise.[23]

If, however, we read Eriugena's exegetical choices from the perspective of universal nature rather than human nature, the turn from protology to eschatology may not be the crucial pivot it has been made out to be. I am swayed by Cavell's Emersonian argument that "the need for a call for change gets expressed as an imperative when what is problematic in your life is not the fact that between alternative choices the right one has become hard to find, but that in the course of your life you have lost your way."[24] Thus, I desire a position that is more embedded or grounded, defined by the fullness of nature rather than dictated by the vanishing point of sin. I no longer think that the exegetical hinge of this grammatical moment in the *Periphyseon* warrants the climactic interpretation it has received. In-stead, it prepares us for what I see as a deeper "conversion." Contrary to standard readings of tensive hexaemeral exegesis whereby irreversible crisis crests in Adam's expulsion from paradise, we should not seek the opposite by latching on to the miraculous change from procession to return in the *Periphyseon*. As if only a last-minute intellectual U-turn can give us the secret key to the correct interpretation of what was, by all accounts, a work of long deliberation and contemplation. When approaching the *Periphyseon* in an Emersonian way, what matters is not that procession morphs into return and apocalyptic crisis is averted but that creation continues its journey home in such a way that, congruent with the divine turn from lamentation to consolation, the entirety of *natura* is reconciled within itself and can thereby assume the force of revelation.

As if to underscore the point that what ultimately matters is *natura*'s self-reconciliation as a condition for its ability to convey revelation, Eriugena's exegesis can display dramatic and even existential overtones. Eriugena no longer conforms to Augustine in this overarching view, focused as he is on restoration as a wholesale return that does not confront or remedy sin but overcomes it by integrating it. In the divine speech that follows the above passage, Eriugena hones in on God's powerful vow "I grieve for him" (*eique condoleo*). Eriugena does something far more consequential than the "U-turn" by nesting the restoration of the divine-human relationship inside of what can truly be seen as nature's ongoing self-revelation. The possibility for conversion emerges only when nature's ongoing self-revelation overcomes the tension between divine accusation and human consciousness of guilt and when God takes pity on the nakedness of Adam by shielding him like Job.

Fed by this tiny spark of divine solidarity, human conversion fans out into the full and frank holistic conversation that *natura* was meant to constitute all along. Here Eriugena's tone acquires all the traits of a frank Emersonian sermon:

> Here we are to understand that the Divine Mercy and infinite Goodness, so ready to forgive and pity us, to sigh over the fall of the Divine image, and in His clemency to condescend unto us and to bear in patience the arrogance of man [*casum diuinae imaginis suspirans, misericorditerque condescendens, hominisque arrogantiam patienter sufferens*] is saying: Now therefore, I behold man driven forth from paradise [*iam de paradiso expulsum hominem uideo*]; formerly blessed, now become wretched; once rich, now needy; once an eternal being, now a temporal; once enjoying everlasting life, now mortal; once wise, now foolish; once a spiritual creature, now an animal; once heavenly, now earthly; once enjoying eternal youth, now growing old; once happy, now sad; once saved, now lost; once the pru-

dent son, now the prodigal; straying from the flock of the heavenly powers I behold him, and I grieve for him (*eique condoleo*). For it was not to this end that he was made: he whom you his neighbors and friends (Job 19:21) now behold driven forth from Paradise into the region of death and misery [*quem uos, o uicini eius et amici, in regionem mortis atque miseriae uidetis nunc de paradiso expulsum*] was formed for the possession of eternal life and blessedness, to consort with the heavenly orders who had adhered to their Creator and remained in everlasting bliss—though a number of them were lost in man's transgression.[25]

Instead of being extrapolated from nature, salvation is, and always has been, fully ensconced within it. At no point does Eriugena invite us to climb a pathway to mystical union with the divine. He simply allows his readers to learn from Adam and Eve by resuming their lives, albeit with a new awareness of the inherent dynamism of those lives. From this frank sermon, articulating the intra-*natura* conversation that constitutes created life in all its brokenness, the vision of return flows naturally.

PART 2

4

Nature as Dispositive Thought in *Schleiermacher's* Speeches on Religion

Introduction

In 1799 a shot ricocheted around the world of German intelligentsia when a young Friedrich Daniel Schleiermacher, barely thirty years old, published his *On Religion: Speeches to Its Cultured Despisers*. Written in the space of two months,[1] these so-called speeches (*Reden*) were never delivered as actual speeches, yet their impact could hardly have been more dramatic. In light of their sudden genesis they have proven remarkably long-lasting. In dramatic effect the *Speeches* seem to anticipate Emerson's "Divinity School Address" of some thirty years later, though, by contrast, the latter's impact seems to have evaporated soon and has not defined its author as a prime religious voice. The *Speeches* would come to define Schleiermacher's religious centrality, kicking off what was to become a long and distinguished theological career that would crest in the voluminous *The Christian Faith* (*Der christliche Glaube* or *Die Glaubenslehre*). This work, a culmination of the entire program that he first unleashed with his *Speeches*, earned him the honorific "father of modern theology," the aptness of which remains uncontested to this day.[2]

Showing more endurance than previously thought, however, Emerson has of late enjoyed a surprising comeback, which affords me the opportunity to put the spotlight back on the "Divinity School Address" and read it more profitably not as a screed or a scandalous pamphlet but a much more subtle, theologically informed text. With the initial scandalous impact of the "Address" having worn off, its relevance need no longer be confined to the event of its delivery alone. When turning to this and other Emersonian texts, I have been keen on gazing through their compact curtness, in an attempt to recalibrate the balance of content and cortex in the interpretation of these texts. Such a recalibration ranks among my goals as I evaluate Emerson's naturalizing thought, and I will return to it in my conclusion.

Unlike the exuberant flair of Emerson's "Address," which marks a stark difference with the equally impassioned yet discursive and, as I will show, dispositive, *Speeches*, Schleiermacher uses a more introspective, meditative writing style.[3] In conformity with this difference of authorial disposition the *Speeches* take us on a probingly mindful, deliberative quest, which makes them hard to put down once begun. Cogent and compelling as they are, their irresistibility is undeniable, even if the reason for that is not always readily apparent. Should we consider their sweeping character a side-effect of their Romantic nature, as the author's intellectual impressionism colors not only his exposition of religion but also, in one and the same sweep, of nature? Furthermore, as the meaning of nature and that of religion reinforce each other in the *Speeches*, what is the reader to make of their respective status, especially when it is not always clear whether Schleiermacher wants to push a clear, delineated argument or wrap his readers in what is at root an enchanting, perhaps even mesmerizing or flirtatious (*schwärmerisch*) exercise?

As per Schleiermacher's stated goal, the *Speeches* have generally been considered an overt attempt at redefining the meaning of religion to a blasé Enlightened audience, aptly named its "cultured despisers." Still, that formal rationale for their coming into being raises more questions than it answers. In addressing religion's cultured despisers, does Schleiermacher, in the aftermath of Kant's epistemological challenges to religion, want to undertake a more methodological course, even if still embryonic, to protest the latter's sacrilegious philosophical coup? That coup consisted in Kant's abrogating any firm human insight into or access to the transcendent by insisting on faith rather than knowledge, while relegating religion essentially to the realm of ethics. God is a postulate of practical reason, and the traditional theological means of divine access through scripture and the church have been deprived of their meaning, leaving Jesus Christ an archetype rather than a mediator. Does Schleiermacher perhaps want to prepare the ground for a new religious perspective, one that no longer situates itself inside rigid institutional structures but is not necessarily constrained within "the limits of reason alone"?[4] If we indeed deem it Schleiermacher's main motive in the *Speeches* to signal interest in a deeper and more radical makeover of religious discourse, there might be further goals to be discerned as well. Does he not also perhaps aim to break through the stale Reformed theological tradition of his day by rekindling the original piety that once ran through its veins and that seems to motivate him more broadly?

Moving the beacons of public debate to accommodate religion's new and freer role, as Schleiermacher might have been trying to do, did not by definition entail the sacrifice of all that came before. In fact, Schleiermacher appears to take past theological concerns rather seriously, even if he aims to channel them into a new frame that can withstand Kant's rational demands but maintains what he deems religion's universal aspiration to actively channel the transcendent. The latter intent seems also to be entailed by the honorific "father of modern theology." While a fuller assessment of this epithet obviously requires us to look at Schleiermacher's entire oeuvre, not only the *Speeches* but also *The Christian Faith*, it remains true that the *Speeches* provide the initial impetus for his modern theological makeover, with his continued edits proving their lasting importance for that project.

I see Schleiermacher's *Speeches* as a relevant kickoff point for the second part of my project on "thinking nature and the nature of thinking." In this second part I will likewise explore the themes of pantheism and selfhood in relation to nature and religion. My aim to assess the crucial relation between nature and religion is what leads me to foreground the *Speeches* and only make passing comments on *The Christian Faith*. Prior to engaging Schleiermacher in the *Speeches*, however, we need to discuss our transition from premodern to modern religious figures. After all, it is on this shift that the book's second part is premised.

Nature's Premodern Guise in Eriugena

To set up the transition to modernity's religious take on nature, let us take a bird's-eye view of Eriugena's dynamic nature, as I analyzed it in Part 1, flanked by accompanying views of Maximus and Augustine. When I first rolled out the idea of dynamic nature, I did so by setting up an axial conversation between Emerson and Eriugena, after which I considered the subtle ways in which dynamic nature interlinked God and self. In my treatment of various premodern authors, this interlinking of God and self generated an inclusive, totalizing sense of nature. Although it was thoroughly impregnated with the divine, the agency of the human self was left remarkably unimpeded, meaning that, despite its fallen state, the Adamic self did not incur the stigma of Icarus-like hubris but remained on track as it continued to unfurl its investigative course through nature. Consequently, nature in Eriugena became marked as a terrain largely beyond the (Adamic) self's, or anyone else's, control.

While I consider Eriugena's view emblematic for the all-encompassing quality of what I called that other, elusive nature, from the start I have depicted dynamic nature in the modern terms of the Emersonian circle. Nesting Eriugena's premodern dynamic nature within the modern Emersonian orbit allowed me to interpret its striving for ever-greater generality as an endemic trait, not a sin-induced flaw or transgressive act. It also allowed me to rule out that Eriugena's stance, deemed heretical several hundreds of years after the fact, was the result of confusion or obfuscation. Taking what is perhaps best called a phenomenological approach, I gradually articulated nature's ever-expanding role as inclusively and expansively divine but for that reason no less moderated and measured by human rationality. This leaves the pantheism charge as one largely misdirected if not irrelevant.

In conformity with the above analysis, any valid theological evaluation of Eriugena should give an account of the status of nature as integrally linking God and self. In the *Periphyseon* divine reason is communicated through the divine Word—by which God creates the world but which, through the incarnation, is also dispensed directly to the human self to be disseminated in redeemed language[5]—and we see that world and W/word(s) have become inseparably linked. Cemented by the *Periphyseon*'s hexaemeral exegesis, the synergic nature of that world-W/word(s) bond is such that even the divide between things or objects, on the one hand, and words or signs, on the other, is no longer insuperable.

Armed with the above vision, we are in a better position to determine where late medieval and modern analyses of premodern dynamic nature may have gone wrong. In their rush to make Eriugena's—or, frankly, any—dynamic nature fit a procrustean mold, with creation a static object at the random disposal of an omnipotent divine subject, earlier analyses generally overlooked the lively interaction of God and self that underlies and animates it. Pursuant to such a reductive view, the *Periphyseon*'s outsized scale and transgressive discourse met with irritation, followed by marginalization and neglect, until the work was entirely circumvented or avoided. The more premodern *natura* defied forensic definition, the more its integral meaning receded. In the end all that was left was an allergic reaction to any nature, deemed pantheistic, whose foreignness was sometimes underlined by ascribing it to alien (read: gnostic) influence.

The ascription of Eriugenian nature to gnosticizing tendencies marks what I see as the double failure of traditional historical-theological analysis. Mistaking Eriugena's efforts to render dynamic nature in livelier, more naturalizing terms for a heterodox attempt, the gnostic reading exacerbates the impression of the irrationality of nature and thereby hardens the paradigm of creation as one that inherently lends itself to rational scrutiny and, by extension, to divine oversight and control. Premodern creation is normatively seen as a channel for undiluted divine inspiration but otherwise depicted as impervious to aspirational human dreams and desires. Put in stark terms, to reduce creation to a conduit of divine omnipotence not only renders nature static and inflexible but, in the final analysis, holds up its passivity as divinely sanctioned and ordained. No wonder that all that was dynamic, powerful, and aspirational about nature in premodernity was increasingly squeezed out of late medieval scholastic views of creation, only to migrate back into nature in modernity, be it under the rubric of Romanticism or pantheism rather than science. The question of course is why nature's rational, scientific organization had to be separated from its poetic voice and inspiration?[6]

In the concrete case of Eriugena, condemnation hit in 1225, more than three centuries after his death, when the *Periphyseon* fell victim to the same suspicion of Aristotelianism that had tainted the reputation of the University of Paris at that time. Linking him next with the heresy of Amalric of Bene, among whose suspect ideas we find not only pantheism but also the eschatological prospect of gender equality, all but guaranteed the work's continued negative reception. The *Periphyseon* eventually landed on the Index of Prohibited Books, where it would remain until that document itself expired. Despite the printed edition by Thomas Gale in 1681—under whose interpretive and Latinized title of *On the Division of Nature* the *Periphyseon* was known until Edouard Jeauneau's reinstatement of the original *On Natures*—it was not until 1938 that Eriugena was rescued from obscurity and perennial suspicion, when Maïeul Cappuyns's monograph unleashed the modern critical study of Eriugena's thought.[7]

A less compressed version of the above development, particularly Eriugena's excision from the broad Christian tradition of nature and creation, prompted my initial search for a new account of dynamic nature in the religious West, even if my evocative approach falls short of rehabilitating Eriugena. Since no historical reconstruction of previous misinterpretations can override the past, rehabilitation is neither needed nor intended. Instead, my analysis in Part 1 aims to lend a voice to dynamic nature. Giving it a voice in current debates makes a world of difference, as deblocking access allows nature's dynamism to flow once again. Rather than designating nature as a contrived alternative to creation, I want to convey it as the unpredictable, wild and wily amalgam of God and self that it is. Reflecting their synergy, and enriching the notion of voice, Part 1 concluded on the note that Eriugena's premodern nature is itself best described as conversation, inasmuch as its empowering quality allows not just for a full partnership but the cocreatorship of God and self.

To substantiate and round out my analysis of dynamic nature in Eriugena (by way of Emerson) in the first chapter, I flanked the premodern part of my analysis with chapters on Maximus and Augustine. Since we know that Eriugena read both of them extensively, this approach allowed me to present Eriugena as less isolated than he is often made out to be. But there is an added advantage as well. Using an updated Eriugenian nature, couched in modern Emersonian terms, as a kind of heuristic yardstick made it possible to read these earlier authors also against the grain of the interpretive vision of a procrustean creation. On such a corrective reading Eriugena's nature, even if still more dynamic, is less out of touch with his peers than his pantheist reputation led us to believe. Hence we discover that what tacitly passes for orthodox creation, and is placed in contrast to Eriugena's nature, need not be considered so restrictive.

Schleiermacher's Spinozism

As the flanking excursions on Maximus and Augustine in Part 1 helped to make clear, Eriugena's nature betrays a dynamic, but not thereby pantheist, quality. Approaching him alongside these intellectual predecessors, rooted as they still are in the early Christian theological tradition that mixes exegesis with philosophical reflection, helps us to strip the lens and charge of pantheism from Eriugena's nature. Yet it is that label that has become affixed to the *Periphyseon*, certainly from the thirteenth century onward. And that same label surfaces in regard to Schleiermacher's thought—and also, if less so, Emerson's—not as an external stain on his reputation but as an intellectual problem with which he chose to wrestle. Hence, it is useful to think through the theme of pantheism and its relation to selfhood as we tackle his *Speeches*.

As Julia Lamm elucidates in her study *The Living God: Schleiermacher's Theological Appropriation of Spinoza*, the contemporaneous response to Schleiermacher's *Speeches*, as well as to *The Christian Faith*, was that they were indeed pantheistic. To this, Schleiermacher replied with almost Eriugenian aplomb, in a letter to F. H. Jacobi in 1818: "Are you better able to conceive of God as a person than as *natura naturans* ("a nature that natures," God as a free cause)?"[8] Clearly then, if not his affiliation with pantheism, then certainly his affiliation with Spinoza was not something about which he appeared worried. As Lamm's study lays out in careful detail, that may have been his attitude more generally. The reason would seem to be that, spurred on by the work of Jacobi, Schleiermacher had worked through some of the thorniest aspects of pantheism even before the *Speeches*. He was thus ready to tackle so-called Spinozism, the phenomenon of Spinoza's widespread religio-cultural influence, head-on, even though he appears never to have read Spinoza. But even at the early date of the *Speeches*, it seems he did not feel the need to fight off suspicions or fear that they might haunt him indirectly, since he had already thought through this charge. His later course in *The Christian Faith* is hence not so much a course correction as a willed and continued development.

Lamm gives us good reasons for branding Schleiermacher a post-Kantian Spinozist, by which she signals that Schleiermacher uses Spinoza to criticize Kant, targeting the latter's dualism between the noumenal and the phenomenal world and denying that the infinite is outside the totality of the finite, even as he conversely seizes on Kant to update Spinoza, mitigating the pantheistic tendencies by pressing on the distinction between the finite and the infinite. In light of Schleiermacher's reconciled attitude referenced above, one might add that he shows himself to be a liberated Spinozist as early as the *Speeches*. He is liberated, that is, from the fear that his continued association with Spinoza would endanger his orthodoxy or sully his theological credentials. His question to Jacobi in 1818 as to whether the latter is better able to conceive of God as a person than as *natura naturans*, therefore, stems from confidence—a confidence that I see present as early as the *Speeches*. When, after the *Speeches* and later in his career, Schleiermacher writes his more conventional dogmatics, *The Christian Faith*, it should not be seen as backtracking or covering up a youthful aberration but as

reflecting a subsequent stage of maturity. To avoid reading it in too teleological a fashion, it needs to be weighed on its own merits.

As Lamm lays things out, when Schleiermacher first enters the discussion about Spinoza's worldview some years before the *Speeches* (i.e., in the
1793–94 period), neo-Spinozism was a rampant, if somewhat amorphous,
cultural influence in Germany. Jacobi's fear was that Spinozism might lead
to fatalism and atheism, while Mendelssohn, with his so-called refined Spinozism, brought it back within the bounds of traditional theism. Lessing
and Herder also had roles of importance here. Lessing, according to Jacobi,
was attracted to Spinoza's nonanthropocentrism, his denial of free will and
his sense of the *One and All*, even though Lessing continued to believe in
divine providence. Herder, by contrast, translated Spinoza's *substance* into
what he called *substantial force*, adding an inherent dynamism by which a
living God could simultaneously replace the stagnation of both deism and
Protestant orthodoxy.[9]

From the above comments we can patch together the attraction of Spinozism for Schleiermacher in particular, in addition to the fact that it was
the fashionable position with which to intellectually engage. First and foremost, it is clear that Spinozism was not regarded as atheism. If we try to
place Schleiermacher in the *longue durée* context of my project of thinking
nature, Schleiermacher's attraction to Spinozism would seem to be fairly
obvious. The encounter with Spinozism wrested him away from the stale
and stagnant creator God of Protestant orthodoxy, which had become as
removed from reality as any elaborate Aristotelian divine *entelechia* was in
the scholastic era. In this, its impact reminds us of the concept of *energeia*
in Maximus, which had the effect of enlivening liturgical protocol. For
Maximus, *energeia* inheres in individual creatures, but for Schleiermacher,
through Spinozism and the influence of Herder, it seems to have accreted
to a veritable life force.

In building her case for a view of Schleiermacher as a post-Kantian Spinozist, Lamm explains how Jacobi had described seven characteristics in
response to Lessing's question, "What is the *spirit* of Spinozism?" Lamm recapitulates these seven characteristics as criteria by which to judge whether
Schleiermacher should be said to be *Spinozist*. Below, I will follow her summary
closely, inasmuch as it allows us to develop a more granular interpretation of
Schleiermacher's worldview.[10] According to Jacobi, *Spinozism* assumes, first of
all, the principle of *a nihilo nihil fit* (nothing comes from nothing), which is

perhaps best explained as the optic negative of the Platonic adage that like is produced from like. This starker and more apodeictic formulation seems to question the doctrine of *creatio ex nihilo* insofar as it widens the gap between God and the world, even though this doctrine had assumed an unassailable position as the stand-alone touchstone of orthodoxy. Second, taking seriously this, perhaps iconoclastic, premise means that in Spinozism there can be no difference between the natural and the supernatural world. Since there can be no transcendent God working on a finite creation from the outside, every finite thing must come from another finite thing, which must have an efficient cause, to stay within rationalist scholastic terminology. This implies, third, that there is only an immanent cause, "eternally unchangeable in itself, which, taken together with all that follows from it, would be One and the Same."[11] It follows, fourth, that there is an infinite regress of (finite) causes, since there is no forensic or extramundane cause, and that, fifth, this immanent cause can have neither intellect nor will since, because "of its transcendental unity and constant absolute infinity, it cannot have any object of thinking or willing."[12] Sixth, the "spirit of Spinozism" must deny the existence of final causes, described by Lamm as extramundane causes willed by an intelligent divine being to bring about some effect in relation to particular events and persons. Therefore, it is, seventh of all, by nature deterministic.

It appears that Jacobi, who had laid all this out in his work *Über die Lehre des Spinoza* ("On Spinoza's Doctrine"), did himself retain a preference for "an intelligent personal first cause of the world,"[13] while we have already seen that Lessing was drawn to Spinoza's nonanthropocentrism. Betraying reservations toward the notion of free will and the concept of a transcendent God, Lessing preferred the idea of the *Hen Kai Pan* or *One and All*. Given the plethora of metaphysical options thus available, Schleiermacher may yet have had his own reasons to turn to Spinoza. Informed by the intellectual climate *du jour*, he appears to have made an intellectual choice that had much to do with his view of Kant as a thinker whose philosophical criticism he felt compelled to accept but whose displacement of religion he considered problematic. Following Lamm's astute observations on this point, I agree that Schleiermacher resonates with many of Jacobi's concerns: in particular, as regards the need for a living God, for a viable realism in the wake of Kant's transcendental realism, and for a credible way to push back against sheer speculative thought. In his own pushback against speculative thought, Lamm sees Schleiermacher shift subtly from Jacobi's assessment of Spinoza as being close to Leibniz to a view in which he reads Spinoza as edging closer to Kant. This allows Schleiermacher to identify Spinoza's *natura naturans* with the living God he seeks, though it is not the personal God demanded by Jacobi. Rather than seeing *natura naturans* as inhabiting rationalism and materialism, as one might expect a typical reading of Spinoza to entail, Schleiermacher's more Kantian reading has Spinoza confronting and opposing both. In this way Schleiermacher's reception of Spinoza is distinctly critical. According to a pattern that Lamm detects, where Schleiermacher disagrees with Jacobi's concerns, we find him defending and affirming Spinoza, as on the points of free will, final causes, and a personal God.[14] Yet, for Schleiermacher, not only does Spinoza seem to stand in opposition to Leibniz, and hence in alliance with Kant, but Spinoza's entire system is attractive because it is ultimately seen as more coherent than that of either Kant, Leibniz, or Jacobi.

By way of summary, let me mention in passing some of the finer judgments that Lamm gives us.[15] She assesses that Schleiermacher is indeed *Spinozist* in terms of Jacobi's criteria, that he is *neo-Spinozist* in terms of his translation of *natura naturata* ("nature that is natured," creation as effect) into an organic web of created powers and forces in line with the inherent immanent cause as eternally changing, and, finally, that he is *Spinozan* inasmuch as he betrays parallels with Spinoza's philosophy, even if he did not know it firsthand. As I indicated earlier, her conclusion leads her to embrace Schleiermacher as a *post-Kantian Spinozist*, a view she hones further by calling attention to, respectively, his organic monism, his ethical determinism, his critical realism, and his nonanthropomorphism. Leaving those categories for what they are and moving on to the *Speeches* now, we know that Schleiermacher is positively disposed toward theology, given that Spinozism is not atheism for him, and also toward cosmology, enamored as he is of Lessing's *One and All*. Most important, however, at least in my own reading of Lamm, is that thanks to Spinoza, Schleiermacher is able to see these fields as connected perspectivally rather than as diametrically opposed or rigidly compartmentalized.

Religion and Cosmos in the *Speeches*

To help us set up a proper debate about the *Speeches*, I want to make a final point about nature and pantheism, a point that touches on the notion of cosmos or *universum* as Schleiermacher employs it in relation to religion.

In his *The Historicity of Nature: Essays on Science and Theology*, Wolfgang Pannenberg references Schleiermacher in a casual comment, on which I want to base some methodological reflections. Letting on that the term *universum*, in the context of speaking about people's "religious encounter with the totality of their being," may be equally extensive of self as it is inclusive of God, Pannenberg remarks that when people experience themselves in the *universum*, "as Schleiermacher said," they generally consider what is beheld in the finite content of that experience revealing.[16] It seems Pannenberg gestures here to a notion of revelation that is neither the Reformers' idea of a direct impact of the divine word nor the religious disclosure that opens up as a result of postmodern deconstruction. Bearing instead a more anthropological imprint, this notion of *universum* as revealing helps to explain why we find Schleiermacher resorting to the cosmos in the midst of giving an apology for religion that relies heavily on human self-consciousness. It also shows an incipient divergence with the later Emerson.

Although I briefly gestured to nature's self-revelation in a phenomenological sense at the end of the Postscript, I have otherwise refrained from deploying revelation as an analytical tool in this book. Since *revelation* is not a term prevalent in premodern Christianity, it was not particularly germane to the discussion in Part 1. Not being Eriugenian, it is also not particularly Emersonian; however, Pannenberg's point about *universum* experienced as revelation is keenly relevant for my reading of Schleiermacher on multiple grounds. The reason for its relevance lies not only in my endorsement of Pannenberg's assertion that revelation is the finite content of the religious encounter with totality—whether that totality is dominated by God or by self is of secondary concern for now—but also in his implicit suggestion that it is a prerogative of religion to express and analyze the meaning of totality, for which revelation is an apposite term.[17] Although there is no reason per se why it could not also be a philosophical prerogative, in the final analysis such is not the case for Pannenberg nor, as he appears to imply, for Schleiermacher. Notwithstanding the latter's wrestling with the challenge posed by Kantian thought, a tension both mediated and assuaged by his affinity with Spinoza, it is clearly by rethinking the depth of his *religious* engagement that Schleiermacher aims to move forward. Taking religion in the sense of the reflection and communication of a comprehensive and holistic worldview, furthermore, Pannenberg seems to gesture at the existence of an organic bridge linking cosmology and theology. In the *Speeches* these are indeed Schleiermacher's twin channels through which he conveys his revelation of totality, irrespective of whether we want to see it in Lamm's nomenclature as Spinozist, neo-Spinozist, or Spinozan.

But more needs to be said. By identifying, and subsequently thematizing, revelation as the finite content of the encounter with the universe or the totality of being, Pannenberg implicitly contracts nature—in contrast to the dynamic holism of Eriugena's meandering *natura*—to a still-life object to be observed and experienced. Divorcing it from any actual divinizing movement and reducing the holism to being itself the content, Schleiermacher's totality, at least on a Pannenberg-type reading, draws the self out of its biblical Adamic contours and accords it the profile of the modern Romantic subject beset by stifling vulnerabilities and individual pressures and torn between seductive attractions and willful objections. Since, with Schleiermacher, we find ourselves embedded in Enlightenment modernity, the question of nature's contours and movement recedes behind that of what animates the self. In conformity with this I cannot but see in Schleiermacher's religious apology a keen desire to unpack the self's inalienable potential, allowing it to freely master, at least in theory, any subject on which it chooses to develop a hold. On this precise point his *Speeches* refer at various times to so-called experts, not just experienced observers but metatheorists, as it were, among whom he doubtless also counts himself. This does not mean, however, that he sees religion as only theirs to analyze. In Schleiermacher's thoroughly modern, as opposed to Eriugena's premodern, world the human self is no longer conditioned to communicate with the divine through the *persona* of the liturgical "I" or use scripture as the communal sieve through which intellectual meaning is filtered before being transmitted. Left to its own devices and metaphysically and ecclesially orphaned, as it were, the self must encounter the totality of being or the universe on its own. This leads the self to call it out as revelation, in the face of whose imprint the weighing in of experts takes second stage. Naming the mystery of the universe as revelatory further underlines the self's ability to arrogate for itself whatever theory it wants to develop. Yet its freedom also comes at a price. Through the self's naming of the totality of being, the observation and experience of that totality or cosmos becomes ever more distanced from the self, as naming hardens into framing.

The notion of framing rather than naming helps us to better understand Schleiermacher's approach to nature inasmuch as it interferes with, and thereby effectively alters, the more embedded (in exegesis and/or liturgy) attempts at natural expression known from premodern practice. As the *Speeches* progress, it seems there is a clear advantage for Schleiermacher to bracket off the religious aspect of the cosmos (even if he does not lift it out completely) in ways that were unavailable to our earlier company of thinkers for whom the divine was always folded in with nature, and there was neither drive nor need to disentangle the two. Schleiermacher's approach to move from naming to framing, while modern and enlightened, also distinguishes him from his Spinozist and Romantic peers, who display few qualms about blurring cosmos and religion completely. Given that Schleiermacher, by contrast, is keen on preserving what he considers religion and, thus, on guarding a line between religion and nature, the question arises where that leaves nature or the cosmos.

Notwithstanding the fact that Emerson scholar Robert Richardson has drawn attention to the affinity between Schleiermacher and Emerson,[18] especially in "The Divinity School Address," I consider Emerson's response to modernity to be rather different. I will elucidate this through periodic asides and excursions in this chapter and the next. Emerson interweaves selfhood and the divine in nature through means other than revelation: he may play with the opposition between self and not-self, for example, and he invokes the curious notion of a double consciousness. Emerson's notion of double consciousness (to be sharply distinguished from W. E. B. Dubois's socially and racially embedded notion of the self looking through the eyes of others) appears in his 1842 essay "The Transcendentalist." The concept is somewhat reminiscent of Eriugena's earlier *duplex theoria*, which for Eriugena facilitates human expression of the divine while maintaining appropriate distance from it. Operating on a horizontal rather than a vertical plane, Emerson distinguishes the life of understanding and that of the soul as simultaneously active but not connected.[19] But revelation, as a device that turns naming into framing, and the absorption into selfhood that it entails would seem of more direct consequence to William James. Meanwhile, something newly encountered in the *Speeches* is not just the self's slow turning toward and thereby attuning itself to the universe, as was the premodern perspectival procedure, but making the cosmos stand squarely before oneself as a framed object to be taken in.

Pannenberg has the right instinct, it seems to me, when he regards the *universum* in Schleiermacher not only as deeply religious but—through Schleiermacher's emphasis on the totality of being and revelation as the finite content of experience—as a possible substitute for religion or, conversely, religion as a substitute for it. Whatever the case may be, the central question before us about Schleiermacher is where that leaves nature as the dynamic concept evoked by my earlier analysis, one that is not only integrative of God and self but ultimately deeply intractable in its holism. What is the fate of nature in Schleiermacher's *Speeches*, in other words, if we conceive of it as an Emersonian circle endemically striving for ever-greater generality? What, in sum, is the future of our cosmic world in Schleiermacher, the role of the self thinking inside of it, and the place of God ordaining it?

Schleiermacher's Cosmos and the Intentionality of Revelation

The young Schleiermacher in the *Speeches*—there are five *in toto*—gives us a layered outlook on religion that makes this early work difficult to gauge and nearly impossible to pin down. His intellectual infatuation with Romanticism may have propelled him in a mystico-religious direction, one that he develops in *The Christian Faith* in a more theological and teleological fashion, while adjusting the *Speeches* through subsequent editions.[20] It is in his revisions of the *Speeches* that Schleiermacher betrays a constant and keen awareness that Kant's oeuvre—not just the principled epistemology grounded in it but also the broader aura of enlightened knowledge radiating from it that seems to thwart direct contact with the divine—poses a serious challenge to the Christianity he knew. Whether or not he is compelled to move away from inherited tradition on account of this Kantian influence, at this early stage it seems that he is, in fact, interested in spreading his views in various directions. Also, while the *Speeches* show him wrestling with how best to meet the Kantian challenge to received Christianity, as Schleiermacher pushes himself outward in trying to express his views on religion, it increasingly appears this challenge acts as an inner catalyst for him to reorient Christian faith and the Christian religion on a nondogmatic, philosophically viable, and critical footing. Whether or not Schleiermacher can overcome Kant even as he appropriates him to the extent that he is attracted to him, Lamm helpfully clarifies that in both cases, that of overcoming and that of critical appropriation, Schleiermacher's thought process likely goes through Spinoza, hence involving the status and religiosity of what we might indeed call the totality of cosmic nature.

A more pernicious challenge following from this than any implication of nonorthodoxy is how Schleiermacher will achieve the required reorientation, if not conversion, of Christianity such that the universe can retain a sense of wholeness. I refer here not to a wholeness that irons out all the layers of his Platonizing discourse or completely absorbs whatever Romantic attitude he may bring to his writing (as some form of perspectivalism will always be *de rigueur* for him); rather, I refer to a unifying vision whose cohesion and integrative power credibly withstands the harshest onset of Kantian criticism as it unfolds along transcendental lines. In short, how can Schleiermacher ensure that God is more than a postulate without relapsing into biblical dogmatism? Seen from this angle, Spinozism cannot but make his task more difficult. To the extent that it disallows compartmentalization, it not only forces Schleiermacher to bridge the Kantian divide of metaphysics and morality but also blurs the distinction between God and creation to the extent that they threaten to be wholly absorbed by each other, and subsequently by the *One and All*.

Laying the groundwork for a fuller response later on, let me reply here that the success of Schleiermacher's *Speeches* pivots on a sophisticated rhetorical strategy to lead his readers onward by dangling before them an uneven choice of "either/or." By this I mean that Schleiermacher attempts to move his readers forward by enforcing a choice and in this way pursues a deeper drive for unity while sidestepping division and fragmentation. Whether he can secure and sufficiently anchor his quest for unity and, especially, whether he can thus make the *Speeches* come full circle—that is, bring his audience around to an acceptance of religion—will need to be evaluated further. The same holds true for the question of what doing so implies not only for the blurring of the line between God and creation but also, as was discussed earlier, for the distinction between nature and religion, and between God and selfhood.

The complexity involved in holding up the spectral duality of "either/or" as a residual rhetorical trope to inch closer to a unifying vision explains why attempts to situate Schleiermacher on a spectrum from pantheism to panentheism are ultimately unsuccessful. As the *Speeches* amply show, Schleiermacher has a sense of cosmos in which the divine looms large or, conversely, a sense of the divine that is profoundly cosmic. Yet, as he presses on with his attempt to persuade his opponents, the cultured despisers of his day, about their need for religion and its larger value, he slowly but inevitably whisks the experience of religion and the divine away from that of the cosmos. In doing so, he erects a fault line wherein Spinoza and, before him, Eriugena merely represent a perspective. As this fault line becomes more pronounced with Schleiermacher firming up the conceptualization of his theology of creation in *The Christian Faith*, and in subsequent editions of the *Speeches*, we can do away with the idea of pantheism altogether. But even in the first edition of the *Speeches*, given his move toward positive religion, it is hard to see how pantheism applies. Far from returning to the theism of Protestant orthodoxy, however, his dynamic universe continues to separate him from traditional views of creation, with his disregard for the Hebrew Bible putting him at a far remove from any of the premodern thinkers discussed in Part 1 here.

Panentheism, meanwhile, pantheism's less heavy-handed alternative, according to which nature is not on par with or a part of God but is instead held by or at rest in Godself, seems equally inapplicable as an analytical tool. The rationale for disqualifying it is similar, inasmuch as Schleiermacher's readiness to lift religion and the divine out of creation and nature or, conversely, distance creation from God, precludes seeing any divine cradle as creation's natural home. While Schleiermacher regularly employs the image of the circle,[21] not unlike Emerson later on, and aims to bend the arc of the *Speeches* to form a unifying vision, the concept of creation actually gives his trajectory linear rather than circular shape. This linear trajectory eventually imposes what we might call the intentionality of Christianity's revelation, meaning that, while the experience of the finite is rooted in the infinite, what this experience means is increasingly determined by Christian overtones as the *Speeches* progress. Once weaned from dynamic nature, the finite's—that is, creation's—reified ontological status rooted in and defined by divine causality prevents it from roaming freely about in a divine matrix and *a fortiori* from riding anything like an Emersonian orbit. Insofar as an organic bond with God is concerned, there is a clear point of no return for Schleiermacher's nature, compelling him to pinpoint creation as theistically caused, even if leaving it orphaned in other ways. This adds another tension to the specter of the "either/or" mentioned before, one that carries increasing weight: *either* life inheres in a divinizing cosmos, which the *Speeches* in their earlier, more Spinozist phase seem to acknowledge, *or* humans inhabit a created world that can meaningfully channel their religious inspiration but whose causal conceptualization depletes the cosmos of its power of providing a home for the divine.

Since Schleiermacher opted for the genre of the speech, it seems he cannot achieve what Augustine was able to do when, in crafting his literal Genesis commentary, he asked the question to whom God was actually speaking when creating, even when humanity had not yet been created. Committed to following the biblical voice of the divine, Augustine left enough room for God to step out of the biblical frame and address hearers or readers directly, not only within the book of Genesis but in creation generally. Schleiermacher's speech lacks Augustine's exegetical double entendre. In Schleiermacher there is only the speaker, whose single voice bears rhetorically down on his readers, whoever they are, pulling them in his preferred direction through the subtle pressure of the "either/or," all the while firming up the underlying religious quest in which he sees them all conjoined. This voice, it seems, belongs to the Romantic subject.

"Whoever they are" is a pertinent comment to insert about the readers of the *Speeches*. After all, Schleiermacher leaves little doubt about his addressed audience. The *Speeches* are addressed to the "cultured despisers" of religion: those who are so enamored of the new intellectual developments inaugurated by Kant and reflected in Spinozism and Romanticism that they are ready to throw the baby of religion out with the bathwater of any delimiting, and hence deemed unreasonable, institutional and doctrinal baggage. Having stipulated, in his treatise from 1793, that religion be confined "within the limits of reason alone," Kant essentially stripped the tradition of word and sacrament, which Schleiermacher held dear, of its enduring power. Religion was instead to be transmitted by reason, which not only counsels against the accretion of tradition but might also encroach on, if not perhaps exclude altogether, the sentiments of the faithful. While the *Speeches* may lack a sense of double entendre, they do seem to adequately reflect the author's inner conflict, his sense of feeling torn between the church that is under attack and the intellectual exhilaration produced by Kant's apposite criticism. As he reaches out to the unconverted, Schleiermacher cannot but have considerable affinity for the opposing camp.

Creation and the Infinite

At this point I want to leave behind what is at best psychological speculation and, taking the rhetorical genre of the speech at face value, analyze instead how it shapes and guides Schleiermacher's dispositive arguments. Over the course of his *Speeches* Schleiermacher consistently endeavors to reach across to his readers, in whose camp he seems partially to include himself. To that end he confers on them a communal identity that is the opposite of where he wants to lead them. In this way he invites them into the conversation but, perhaps more important, also prepares them for the slow conversion through the choice of "either/or."

This is how, early in the "First Speech," the "Apology" for religion, he sets in motion the arc of his unifying quest:

> You have succeeded in making your earthly lives so rich and many-sided that you no longer need the eternal, *and after having created a universe for yourselves, you are spared from thinking of that which created you* [*und nachdem Ihr Euch selbst ein Universum geschaffen habt, seid Ihr überhoben an dasjenige zu denken, welches Euch schuf*]. . . .
>
> All this I know and *am nevertheless convinced to speak by an inner and irresistible necessity that divinely rules me* [*von einer innern und unwiderstehlichen Notwendigkeit, die mich göttlich beherrscht, gedrungen zu reden*], and cannot retract my invitation that you especially should listen to me.[22]

The above quotations (excerpted from what is a more drawn-out passage) contain two specific points to which I want to call attention. First, Schleiermacher echoes here what is at heart a very traditional Christian apologetic point, going back at least as far as the days of Justin Martyr. Whereas Justin set up a choice between worshipping gods that are humanly made (i.e., idols) and worshipping the God who made the very worshippers,[23] Schleiermacher's argument is more conceptual, pivoting on the choice between divine creation versus human self-formation. On the one hand, there is the universe that the cultured despisers have fashioned for themselves; on the other hand, there is Schleiermacher's counterview that the universe, anonymously mysterious but amorphous for now, has created them—that is, the very people who so proudly harp on their own self-fashioning. As if embedded in a deeper cultural matrix, Schleiermacher presents this choice—between belief in a divine universe and seeing the world as ours to fashion—as emerging at a critical juncture in time. He thereby lends his readers' impending choice an urgency not witnessed in any of our premodern thinkers but not unlike the crisis mode evoked in Emerson's "Divinity School Address."

Insofar as Schleiermacher's presentation of this critical juncture reflects his cultural moment—but also, as we will see, his pivotal hermeneutical turn—it should be considered relevant for his way of "thinking nature." Or, perhaps, it would be more accurate, in light of the broad Kantian challenge and his defense of the Christian religion, to call it his "reconceiving revelation." The genre of speech (*Rede*), through which Schleiermacher impresses on his readers the need to choose, betrays, despite the increasing starkness of the choice, a porousness to outside pressures, including those potentially resistant to his trajectory of thought. In line with this it seems that in his first speech Schleiermacher's arrogation of voice does not only channel what he has to say but emblematizes a protean authority—rational, individual, persuasive, but also prophetic. With this authoritative voice he presses the choice on them as the central dilemma of the age.

This brings me to my second point. In spite of the gulf that separates writer and readers and provides the impetus for Schleiermacher's apologetic course, we should not overlook the underlying bond that connects him with the cultured despisers of his day. This inchoate, half-articulated bond—of cultural elite and intellectual self-worth, an amalgam of Kantianism, Romanticism, and Spinozism mixed with traces of traditional Reformed piety—allows Schleiermacher not only to push the despisers, at least rhetorically, further than they might be ready to go but also to push himself further than they may have expected him to go. An example of this is when he transitions from the rational to the prophetic mode. In what seems to be a kind of intellectual arm wrestling, Schleiermacher challenges the despisers to reconsider their stance by accepting that his invitation, extended with considerable rhetorical flair, stems *au fond* from a deep inner divine necessity.

As a further inference from this assumed bond, it seems Schleiermacher would like his efforts at persuasion to be followed by a more direct result: the synchronic unity between the minds of his readers and that of the author. Such unity would signal at least the beginning of their conversion but also a deeper recognition of their joint belonging in or intended return to a divinely ordained universe as their home. If we can agree that Schleiermacher, despite his sometimes blistering rhetoric, is nevertheless keen on avoiding a collision course, and see his polemical opening gambit as a proleptic aspirational move, it should be clear that he takes a big risk. After all, it is fundamentally unclear whether he can forge the religious commitment for and with his readers that he so desires and, bringing his entire readership around to his position, actually bring his speeches full circle. It is especially unclear whether the assumed bond with the cultured despisers, on which his *Speeches* are premised, will hold given his periodic predilection for a prophetic role.

If we summarize how Schleiermacher's dispositive argument unfolds so far, we get the following. Early on in the "First Speech," he needles the cultured despisers for having created a universe for themselves, thereby forgoing the difficult task of thinking of a universe that has created them. He counters their intellectual patricide—and note that while he accuses them of having given up on the universe as having created them, he does not discuss the experience of that universe in terms of Pannenberg's sense of encountering a totality of meaning—by inserting himself as the divinely sanctioned ambassador who alone can deploy their intrinsic ties with the creating universe. It is in this vein that we find him stating, "All this I know and am nevertheless convinced to speak by an inner and irresistible necessity that divinely rules me." Schleiermacher retains remarkable rhetorical poise throughout, as he projects that his readers will accept his ambassadorial authority.

A bit later, Schleiermacher references how much the cultured despisers loathe the role of experts and abdicates his own expertise to more fully align with them. In place of his own expertise he thus appropriates the language of divine command when he asserts that the "inner, irresistible necessity of my nature; it is a divine calling; it is that which determines my place in the universe [*es ist das was meine Stelle im Universum bestimmt*] and makes me the being that I am."[24] He paints a picture wherein the divine commands him and, by invitation, his readers. Reinforcing the inchoate bond between himself as the writer and the readers of the *Speeches*, Schleiermacher makes himself an inner instantiation of divine order in that through the inner necessity of his nature, God not only determines his place in the universe but also defines who his and everyone's human "I" most deeply is. In tandem these points show how religion *qua* revelation is indeed the alpha and the omega for him. "Thinking of that which created you," as he wants the cultured despisers to do, is in the final analysis for Schleiermacher not, as we saw in Emerson's *Nature*, an invitation aimed at opening up nature as the place where one makes one's home—and hence fitting for God to reside. Rather, as if carefully threading a needle, Schleiermacher charts a path forward to God through the universe's two eternally opposing forces: nature's attraction that pulls us in, on the one hand, and the expansion of the active and living self that pushes us outward, on the other.[25] Navigating between cosmic mystery and human desire, so to speak, Schleiermacher must persuade religion's despisers that there is indeed a "common bond of consciousness" that surrounds and envelops both,[26] making all humans in the face of totality or eternity ultimately alike. A disclaimer soon follows, however. Rather than relying on his readers to find their individual equilibrium through self-formation, Schleiermacher counsels that when such an equilibrium is found, given that we are all divinely ordained in our place, "it is generally only the magic of nature playing with the ideals of humanity"; it might not even be desirable, as "the final purpose of nature would be wholly thwarted."[27]

Clearly then, some forensic expertise is still needed to guide the audience of the *Speeches* to Schleiermacher's desired end, though it can no longer be that expertise which Schleiermacher renounces when initiating his debate with the cultured despisers. Yet there is an alternative kind of expertise available. This expertise is not extrapolated from knowledge that is the terrain of academics and other scientific professionals; rather, it lies inherent in the religious enterprise itself, however one defines it, and is hence of an experienced nature: "Therefore the deity sends . . . mediators . . . who strive to awaken the slumbering seed of a better humanity, to ignite love to the Most High, to transform the common life into something higher."[28] What is most urgently needed, then—for do recall that Schleiermacher has placed himself and his readers on a critical juncture—is for the expertise of the divinely sent mediators to fuse with the experience of contemporary religious believers. Rather than interpreting this fusion of horizons as an absorption into the fullness of a divinely ruled universe, as would have been the premodern solution, Schleiermacher proceeds by charting a course that is an emphatically modern anthropological one: "May it yet happen that this office of mediator should cease and the priesthood of humanity receive lovelier definition!"[29] From here on, it seems we are dealing with a full-blown apology, with Schleiermacher chiseling and honing the concept of religion as something that is not only decisively different from morality and metaphysics but that resonates with innate human experience and plumbs the depths of human self-consciousness.

Harking back to something inside of him to which he bears witness, Schleiermacher wants to reveal "where this holy instinct lies concealed"[30] and go to "the innermost depths" from which religion "first addresses the [i.e., his] mind (*das Gemüth*)."[31] Through this return to the "innermost

depths," and with the aid and support of the religious mediators, he hopes
to touch that innate capacity in his readers in order to show them experien-
tially that capacity of humanity from which religion proceeds. As the concept of religion becomes ever more pronounced, however, the importance
of the universe as humanity's natural home, jointly housing God and the
self, decreases proportionately.

As the *Speeches* transition from a cosmic to a more anthropocentric discourse, it is as if Schleiermacher hermeneutically stands in for all of humanity, which in the reconfigured unifying quest increasingly needs to bend its
finite mind toward the eternal intuition of the infinite. Self-transcendence
will remain a central theme in the *Speeches*; meanwhile, not only does the
cosmos recede from the center of the debate, but it also no longer delimits
the debate's contours. Meanwhile, religion comes to mean a certain readiness for the infinite to break in. Following on the heels of his rejection of
Kant's religion within the limits of reason alone, and a far cry from Emerson's approach, Schleiermacher anticipates the "essence of Christianity"
motif, in the wide range that it displays from Feuerbach to Harnack. In
that context the cosmos is not to be seen as the excluded middle so much
as the discarded home, its role now redundant, for it is imbricated in the
colloquy between the self and the intuited infinite. Especially in the "Second Speech," we see a clear divergence from natural religion, to be taken
up later in the "Fifth Speech." As he does so in the "Second Speech," he
tells us that he only leads us to the "outermost forecourt" of "external nature," acknowledging that many consider it the "innermost sanctuary of
religion." The fear and joyous pleasure associated with the material forces
of the earth and with corporeal nature may well seem to prepare us for religion, but they do not give us "the first intuition of the world and its spirit
[*die erste Anschauung der Welt und ihres Geistes*]," as "these sentiments themselves are not religion."[32]

In other words, the hermeneutical clarity that Schleiermacher gains by
zooming in directly on the roles of God and religion seems to leave the
cosmos behind. But not only that. The more Schleiermacher takes religion
out of its cosmic home, his concomitant losses are twofold: first, the more
he must also let go of the cultural amalgam that binds him to his readers;
and second, the more he loses touch with the universe—not just in the
sense of revelation, that is, the totality of meaning that humanity's finite
encounter with it represents but also the actual world of nature. Losing the

actual world of nature becomes the concrete price for the reductive clarity of modern religious scrutiny. Rather than thinking about modernity in terms of disenchantment and Enlightenment, what we see in Schleiermacher is that the hermeneutical and the scientific dispositions begin to move along parallel tracks that no longer allow for encounter.

What has been a brewing tension between nature as religious medium and religion as constrained by its "natural" mediation crests in the "Fifth Speech," from whose quandary, in turn, it only seems a small step to *The Christian Faith*. In the "Fifth Speech" Schleiermacher presents his readers with a stark choice between a total absorption into natural religion, which he rejects, and an express commitment to revealed religion, with "revealed" honed in terms of an acquired intellectual tradition that has received cultural and ecclesial codification, one that would incorporate the intuition of the infinite that marks his overarching quest. Most important, though, here we see the evoked duality of the "either/or" of accepting or rejecting religion drive a wedge into Schleiermacher's rhetoric of a common bond; this tensive duality, instead, blossoms out into total separation.[33]

Unsurprisingly at this point, Schleiermacher phrases the quandary I just referenced in terms of a stark opposition to the adherents of natural religion:

> The essence of natural religion actually consists wholly in the negation of everything positive and characteristic in religion and in the most violent polemic against it. Thus natural religion is also the worthy product of an age whose hobbyhorse was a lamentable generality and an empty sobriety, which, more than everything else, works against true cultivation in all things. There are two things that they especially hate: They do not want to begin with anything extraordinary or incomprehensible, and whatever they might be and do is in no way supposed to smack of a school. This is the decadence you find in all arts and sciences; it has also penetrated religion, and its result is this contentless, formless thing. They would like to be indigenous and self-taught in religion, but they have only the crude and uncultivated part of these qualities; they have neither the power nor the will to bring forth the unique.[34]

Hereafter, his rhetoric will lead him away from a mystery-filled cosmos and forward to *The Christian Faith* and a teleological restyling of theology. This theology is focused on the individual and brings out the intensifying bond between the personal believer and the mediator Christ as the grace-filled savior who embodies the teleological religion of Christianity.

Yet by making his choice in favor of the positive religion of Christianity, Schleiermacher also institutes a religious taxonomy that not only hierarchizes non-Christian religions into submission, so to speak, but also rearranges the dynamic within Christianity itself. He assigns the Old Testament, for example, a markedly lower status, reducing it to a narrative preamble to, rather than the site of, the meaning of creation and nature. Whatever remains of nature's evocative power becomes compressed into the thin notion of formal causation. Oddly, the way in which the ambit of nature and creation gives way to the ambit of (overt Christian) revelation by the "Fifth Speech" implies an anthropological contraction of salvation that in some ways gainsays the mystery-laden notion of intuiting the infinite that otherwise characterizes the *Speeches* and has made them so famous.[35] Thus it appears that Schleiermacher is slated on a path that, via the dismissal of natural religion, veers toward the excarnation of nature.[36] After all, if natural religion is summarily dismissed, what is left for nature to represent and do?

Divine Causality and the Excarnation of Nature

In his recent book *Die Macht des Heiligen*,[37] Hans Joas references the problem of natural religion in the eighteenth and nineteenth centuries. Joas, who will be an interlocutor also in the next chapter, on William James, is keen on developing a history of religions that is not merely the opposite of theology but takes religious commitment seriously and acknowledges it as a viable way to account for and deal with religion. To that end he seeks a method that allows for crisp analysis without the implication that one thereby sacrifice the possibility of personal commitment to any one religion. In the context of Joas's focus on Western culture, such religious commitment evidently pertains especially to Christianity.

Following his discussion of David Hume's *The Natural History of Religion* (1757) as representing an attempt at an empirical, secular, universal history of religion,[38] Joas explains that its reception in Germany had a rather different outcome than one might have expected. To the extent that German thinkers like Hamann and his student Johann Herder received Hume, they used their responses to Hume to guide their own way to countervailing what they regarded as the flattening out of Enlightenment Christianity.[39] Given that they considered Enlightenment rationality to blame for turning Christianity into the universal religion that was Hume's ideal, their responses allowed them to come to Christianity's defense. In this way they foreshadowed Schleiermacher's *Speeches*.

Joas is interested in mapping out a genealogy of religion of his own that is meant to counter Nietzsche's more critical if not plainly destructive impulse, that is, destructive of Christianity. Joas concludes from the German reception of Hume that an empirical history of religion à la Hume may be possible but is bound to find itself in a field of tension between religious criticism (*Religionskritik*), on the one hand, and apology, on the other.[40] This leads me to the following set of inferences. As Schleiermacher's "Apology" deployed the rhetorical strategy of "either/or" to forge unity between himself and his audience, whom he considered *au fond* engaged in a common quest with him, we saw this quest break down in the "Fifth Speech" and result in a stark plea for revealed Christianity. Yet if we elevate them to the more abstract, historiographical level of Joas's projected new genealogy, the *Speeches* occupy a place on a spectrum whose poles represent another duality, a different "either/or." In a slight rephrasing of Joas's terms it seems that the *Speeches* find themselves in a field of tension between what one might call the "critical advancement of religion" (rather than religious criticism/*Religionskritik*), on the one hand, and apology, on the other. This apology, however, not only defends religion but, in the process, materially equates it with received Christianity. Here he is heir to Herder.

Trying to position nature or creation inside this field of tension now, as we see Schleiermacher oscillate between a critical reception of religion and his final apology of traditional Christianity, I want to return to two observations I made earlier. The first is that although Schleiermacher references the universe, and in later redactions does so with a special focus on the intuiting of the infinite, he shows remarkably little interest in the ties that bind the universe to Genesis. That is, as I see it, he is uninterested in the connection between "universe" and the materiality of creation as continued in and primarily accessible through the reading of scripture.[41] The exclusion of Genesis shows him to be a thinker who predates the stark theological choice we now (post-Barth) face, sketched in Chapter 1—namely, treating creation either in terms of biblical stewardship or of postmodern thrownness. In the Schleiermacherian paradigm distilled here, creation is first and foremost defined by finitude as the effect of divine infinity, and hence by causality, to which any scriptural legibility takes second place. Interestingly,

Schleiermacher does not himself acknowledge or even sense this as a departure from an earlier tradition to which he otherwise seems to want to pledge increasing allegiance. But in this he is part of a greater slow-moving process of fading out of loyalty to biblical truth given that, in modernity, the center of gravity for Christianity has shifted from exegesis to the philosophical justification of Christianity's tenets and, thereafter, the legitimacy of religious belief. From another angle awareness of this background allows us to bring to the current dilemma the knowledge that creation in Schleiermacher and the pre-Barthian tradition is indeed to be seen as firmly Christian, even if it is not overtly biblical. Some of the features of Schleiermacher's universe are better understood precisely by factoring in this distinction.

For my second, related point I want to invoke the following analogy: just as creation is stripped of its biblical wrapping in Schleiermacher and the relationship between creator and creation thereby contracted to causality, so it seems that religion is undone of its cosmic overtones and as a result contracted to the feeling or consciousness of dependence, Schleiermacher's famous *schlechthinniges Abhängigkeitsgefühl* from *The Christian Faith*. Or, to put the analogy slightly differently: just as in hindsight—and I do not want to downplay the retroactive aspect at work here in my analysis—scripture is conceived and subsequently left behind as the linguistic shell of creation, so the cosmos is in hindsight conceived and subsequently left behind as the shell of religion in the *Speeches*. Given Joas's desire to construct a different genealogy of religion that satisfies his desired universal science of religion, which does not depend on the exclusion of personal religious commitment, one wonders if Schleiermacher's *Speeches* could have given us a more complete natural religion if, in a thicker move and one of greater hermeneutical fullness, he had kept the cosmic shell intact and focused less on causation and more on cosmic wholeness and integration. Integration could have been achieved if the finite intuition of the infinite as the exclusionary vertical quest had not fully merged, through the feeling of dependence, with revealed Christianity in *The Christian Faith*, but had instead retained sufficient latitude of its own.[42] Such was not the case, however, as Schleiermacher's *Speeches* developed in the way they did, that is, as he redacted them and developed his thought further in *The Christian Faith*. In the face of the vertically expressed relation of God and human creatures in the later Schleiermacher, the possibility of an Emersonian orbit must recede.

Nevertheless, it is a tantalizing task to try to find some aspects of what a thicker, more cosmic Schleiermacherian hermeneutic might have looked like. While it can obviously not be reconstructed in its entirety, there are building blocks strewn throughout the *Speeches*, and I want to reveal a few of these. In doing so, I am guided by what I see as an affinity between Emerson and Schleiermacher that differs from what was observed by Richardson—namely, a sense that what makes nature tick in Emerson is akin to what in Schleiermacher makes religion work. Both involve and revolve primarily around the divine, and both strongly influence human nature. Instead of a parallel, however, in my reading we have here a case of triangulation. Whereas nature for Emerson is suspended between God and self, for Schleiermacher it is religion or God that is suspended between the universe and the self. Thus, he says in the "Second Speech": "All these feelings [compassion with injustice in the universe, etc.] are religion, and likewise all others in which the universe is one pole and your own self [*Euer eignes Ich*] is somehow the other pole between which consciousness [*das Gemüt*] hovers."[43] But whereas Emerson and Eriugena are willing to contemplate the omneity of nature, both seeing it as their primary challenge, for Schleiermacher religion is poised to help ward off the chaotic darkness of such an indivisible universe, and contemplation of such a universe belongs only to the unrefined person, leading one to blind chance if not idolatry.[44] It is as if religion is first the crutch of the universe, aiding helpless and hapless humans, before it replaces the universe altogether. In both stages the universe's role is one of fleshing out divine teleology.

If it is an accurate assessment to say that what we find in Emerson is the triadic model of self-nature-God and in Schleiermacher that of self-religion-universe, what does this entail for the status and meaning of the universe in Schleiermacher? While religion in the mature Schleiermacher is defined as the feeling of absolute dependence, in the *Speeches* the universe is not yet put firmly in its demoted place and is still eager at times to trade places with God. In the *Speeches* God sometimes hides behind the universe, in the sense that a kind of intuition picked up in the universe may act as a pointer to God, while at other times the universe is itself pregnant with the divine, the pointer more like an emanation. In both cases, however, there is an impending sense that if and when religion is put firmly in place through right intuition or feeling for God, the universe will become either sidelined or redundant, its mission inevitably considered fulfilled. Hence my comment above about the cosmos as a shell for Schleiermacherian religion.

There is an additional difference here with Emerson on the notion of selfhood that is worth mentioning. Inasmuch as religion in Schleiermacher speaks to the *Ich* rather than implicating a more anonymous, collectivist, amorphous selfhood, there is a need for the *Ich* as self to always be the subject of whatever gazing or intuiting is going on, despite the passive disposition of receptivity with which it otherwise needs to intuit the infinite. Emerson, by contrast, can also divide the world into me and not-me, but his less personal or individualized selfhood allows him to have both a subject side and an object side to it, simultaneously or alternately.

This leads me to two further differences. First, there is a strong sense in Schleiermacher that the *Ich* must be fundamentally other than God/universe, as Schleiermacher's more orthodox religious outlook makes him err on the side of distinction and difference rather than identity or sameness. In my view Schleiermacher's adherence to an orthodox Christian notion of creation comes out more in his unquestioned acceptance of the dividing line between creator and creature (therefore, the human self) than in any fear of pantheism. From this it follows, second, that divine transcendence comes at the price of human transcendence or transcendentalism. Thus, it is true for Emerson, but not for Schleiermacher, that one can look at nature's ball (as in the opening of "Circles": "Nature centers into balls") both from the inside and the outside. As an aside, the constant use of the image of the circle in the *Speeches* is noticeable, but they always seem to be closed circles,[45] in contrast to Emerson's notion in "Circles" that when one has walked the circumference of one circle, it gives birth to a new one of greater generality.

Although the point of the fixity of the I (the conscious *Ich*, rather than a self that can be simultaneously subject and object) may not seem to amount to much, it begins to put distance between Schleiermacherian religion and Emersonian nature, which I set up above as being similar in their mediating role between selfhood and God. While there may increasingly be a subjective aspect to *Anschauung* (intuition) in Schleiermacher, which swings religion ever more to the side of the individual *Ich* who intuits, I want to bring the universe back into the discussion at this point by zooming in on the role of causality. I do so to flesh out what is entailed by Schleiermacher's view of humanity and the world as caused by and depending on God rather than as pregnant with the divine. In doing so, I am guided by comments from Robert M. Adams wherein he observes a similar construal of causality in premodern thinkers, among them Thomas Aquinas.

In his article "Faith and Religious Knowledge" in the *Companion to Schleiermacher*,[46] Adams refutes the idea that for Schleiermacher religion is only about experience or consciousness. To counter that notion, he points to religion's implicit intentionality, in Schleiermacher, or reference to a being much greater than ourselves. Wayne Proudfoot famously accused Schleiermacher of trying to have it both ways: wanting to be both intentional (that is, developing a mental concept) and immediate (that is, forgoing such a concept).[47] In contradistinction to this position, Adams prefers to see Schleiermacher's religious consciousness as preconceptual or independent of concepts but with an implicit intentionality. When Schleiermacher states that "intuition of the universe . . . is the highest and most universal formula of religion,"[48] for Adams this means that the objective genitive ("of the universe") is indicative of an object, hence intentional or referential; yet that object is not the universe but rather "the universe and the relationship of the human being to it." This more inclusive or well-rounded object is, for Schleiermacher, also the object of metaphysics and morality. Schleiermacher concludes: "Thus to accept everything individual as a part of the whole and everything limited as a presentation (*Darstellung*) of the infinite is religion."[49] Schleiermacher concludes this foray with the following, leaving off where Eriugena and Emerson would seem to continue: "But whatever would go beyond that and penetrate deeper into the nature and substance of the whole is no longer religion, and will, if it still wants to be regarded as such, inevitably sink back into empty mythology."[50]

In the second edition of the *Speeches* from 1806, *intuition* becomes more theoretical and scientific for Schleiermacher and is replaced by *feeling*, which tilts more inward than outward. This ensures that the primary religious consciousness now equals a kind of self-consciousness. But for Adams there is still an implicit intentionality in Schleiermacher's words: it is "the one and all of religion to feel everything that moves us in feeling, in its highest unity, as one and the same."[51] In *The Christian Faith*, then, piety or personal religiousness is the feeling of absolute dependence—that is, the fact "that we are conscious of ourselves as absolutely dependent, or, equivalently, as in relation with God." Schleiermacher goes on: "the whence that is implied (*mitgesetzt*) in this self-consciousness . . . is to be designated by the expression 'God.'"[52] Based on this, Adams sees that "God," as the referent in Schleiermacher, gets its content from reflection on the *feeling*, not from any knowledge (of God) that is prior to the feeling, hence not from tradition. It is not true, then, in unqualified fashion that God is the intended object of essential religious consciousness, which, rather, remains self-consciousness for Schleiermacher. The whence of absolute dependence is thus not given in the feeling but inferred from the description or interpretation of the essential religious consciousness as a feeling of absolute dependence.

But where does this leave causality? From his 1799 *Speeches* to the 1830 *Christian Faith*, as I keep following Adams, Schleiermacher insists that religious consciousness is so only insofar as it is consciousness of our being effected by something other than ourselves. This may further explain his reticence, even skepticism, about listing attributes of God, like omniscience, insofar as these do not give us immediate insight into God's character but rather relate back to divine causality, since that is what the feeling of absolute dependence instills in humanity. In other words, God can only be accessed through causality, not by going beyond the concept of God as cause to experience Godself through his attributes. As if to bolster the legitimacy of Schleiermacher's notion of causality, making it less thin and formal by embedding it in theological tradition, Adams observes a connection on this point between Schleiermacher and Aquinas, Maimonides, and Albertus Magnus. While all three authors work with causality but also with negative theology, Adams selects Albert for his comparison. The fact that Albert prefers causality to the way of eminence or negation makes him stand out from Maimonides and Aquinas.

My problem with Adams here is that the connection he establishes with the premodern scholasticism of Albert and others, however Neoplatonic in coloration and therein providing an analogy, seems premised on the use of history as an apologetic mechanism to deflect the essential quandary of Schleiermacher's position. Indeed, Schleiermacher has no Reformed notion of revelation as the direct experience of the Divine Word, but does that justify attributing a preconceptual or implicit intentionality to him that is then somehow bolstered by deeming it rooted in premodern thought? By way of alternative, it seems we may be better off proposing the reverse. Rather than tracing Schleiermacher's causality back to Albert in what is in essence a nostalgic move, a retreat before the potentially radical effects of modernity, it is more sensible to see Schleiermacher's fascination with divine causality as a kind of overblown *reductio ad essentiam*, betokening a holdout of a *Romantic* view. What I mean is that the thin line of causality lends itself to being overread and oversymbolized, not unlike how we see Adams press on it to make it a stand-in not only for the relationship of God and the world within which humanity has its home but also, more broadly, for a secure worldview in which the triangle of God, world, and humanity remains integrally protected, even if in formal—that is, excarnate—fashion. Insofar as, in Schleiermacher, causality *qua* religion's essence is all that is left, not only is there no further reduction possible, but it seems there is *a fortiori* no longer a possibility to reinflate causality with the intracosmic plenitude that characterized the freer, more intimate, and more organic bidirectional interaction between Creator and creation of our premodern interlocutors. Put differently, Schleiermacher's religious universe simply lacks the accordion-like structure of Emerson's nature.

Divine causality can effectively communicate the feeling of absolute dependence, acting as a kind of lens that yields more hermeneutical clarity. Yet it also risks channeling this dependence in a way that presents us with an impoverished universe, one whose religious resonances have become actively suppressed, given that Schleiermacher seems to have channeled religion in a different way. Questions remain, however. Must it indeed be the case that, corresponding to the way in which Schleiermacher fills out the particular God-consciousness of Christianity through his focus on Christ as mediator alongside but also superior to that of other religions, the universe must cease to be formative in matters of faith, since it, not unlike scripture, is no longer itself considered revealing of the divine? This is how Schleiermacher's development from the *Speeches* to *The Christian Faith* can be read, leading us to Barth, whose contraction back to the exclusivity of revelation as scriptural is not just a reaction but also a logical, if paradoxical, corollary.

Or, are there perhaps traces whereby we can see Schleiermacher's cosmos remain a driving force, and does it, for that reason, make sense for Schleiermacher to remain vigilant and push back against its encroaching power, given his commitment in the "Fifth Speech" to revealed religion? Stepping back from the innate force with which modern revelation tends to turn naming into framing, we do well here to heed Schleiermacher's comment in the "Second Speech": "If you put yourselves on the highest standpoint of metaphysics and morals, you will find that both have the same object as religion, namely the universe and the relationship of humanity to it."[53] While Schleiermacher is known mostly for his distinction of religion from metaphysics and morality, what may provide continuity with premodern theological thought is not the thin strand of causality distilled as the essence of religion but the joint and lasting relationship to the cosmos by which, in an earlier phase, all three (God, humanity, world; but also, religion, morality, metaphysics) were bound together in an insuppressible bond. In the twentieth-century, post-Barthian way of reading Schleiermacher that seems to have become *de rigueur*, despite their evident differences, Schleiermacher's separation of religion from metaphysics and morality has come to underline the separateness of religion from any kind of cosmic overtones, as is further brought out by Barth's vehement opposition to natural theology. We should not, however, take Schleiermacher's concentration on causality as a *fait accompli* but probe whether such a reading does not needlessly delimit the possible resonances of his hermeneutical stance, reining in nature as well as scripture. Certainly William James, to whom we turn next, saw a possibility for richer relations among religion, morality, and the cosmos, to the point of considering them to be scientific.

William James and the Science of Religious Selfhood

Schleiermacher, Emerson, James

In the previous chapter our focus was on the *Speeches on Religion to Its Cultured Despisers*, by Friedrich Schleiermacher. Schleiermacher is generally seen as the religious thinker on experience, but as my chapter on him tried to make clear, over the course of the *Speeches* we see the "religious experience" he thematizes undergo a slow change in terms of both quality and identity. It moves from affinity with the cosmos and a largely immanent deity to a more crisply focused dependence on the traditional transcendent God of Christianity, who manifests and reveals himself teleologically through humanity's absolute dependence on him. In terms of my project of "thinking nature," this means that the position of the cosmos as the divine's natural home harboring and channeling divine presence and authority recedes, if not erodes, over the course of the *Speeches*. Nature gradually morphs from being the premier canvas of divine manifestation, expressing its radiant luster, to being the targeted object of divine activity, the finite product of divine causality.

Whereas Schleiermacher's introduction of and attention to the category of experience has led some scholars to detect a connection with Emerson and transcendentalism,[1] I reject such a link on the grounds that it is too preoccupied with funneling Schleiermacherian thought into an experience-focused, prefabricated Emersonian mold. Also, it seems to me that we do injustice to the creative richness of *Emersonian* thought if we let the artificial, that is, *ex post* character of Schleiermacher reception determine the mode of comparison. By compressing a curated selection of Emersonian ideas on experience into such a tidy rubric, we risk skewing them to fit more conventional theological categories. As I have tried to argue throughout, it is precisely these that Emerson defies.

I underlined my disagreement by demonstrating that, if we put the two partners in this comparison on a more equal footing, Emerson is crucially different from Schleiermacher. The more we sharpen our focus to zoom in on notions of nature and creation, the more we discover their differences. To illustrate these differences, I drew out points I first made in Chapter 1, where, analyzing Emerson's nature as pervasively dynamic, I described the role of nature for Emerson as accordion-like in that it can contract to be contained in selfhood in an instant, only to then fan out again to encompass all of reality. As the accordion image led me to conclude, when the attention we pay to nature in our reading of Emerson is either overly granular or overly sweeping, and, especially, when it overlooks temporality as a force inherently pulsating Emersonian onwardness, we end up with results that are inevitably distorted to the extent that they are untethered and adrift in a broader historical, but uncontextualized, framework. Such is the case when we draw Emerson and Schleiermacher together without adequately anchoring each thinker in their own wider ambiance and without sufficiently texturizing their individual positions.

Moving beyond Schleiermacher and his potential relation to Emerson, I want to commence the discussion in this chapter by stating that it is William James (1842–1910), rather than Emerson, whom we may productively link to Schleiermacher. Doing so will not only help us to clarify their respective positions; it will also prod us to shed more light on their shared difference with Emerson, which I see lodged in their more prominent, even if also more abstract and ultimately cosmically isolated, conceptions of selfhood. Once this shared difference is clarified, I hope that the observed link between them can further elucidate Emerson's position, even if it will do so largely obliquely.

But wherein do I see the actual connection between Schleiermacher and William James? And, moving into a more extensive discussion of James, what does it entail? And, especially, how can we bring their connection to bear on my larger theme—the link between "thinking nature" and the nature of thinking? Grappling with these questions, I will, in my reading of William James, pay attention to the position of Charles Taylor.[2] Somewhat surprisingly, in recent years Taylor has singled out William James as the author who offers the most promising paradigm for the study of religion in our contemporary age, which, as is widely known, he has subsequently analyzed as a secular one. Taylor's discussion has potential consequences, therefore, for how best to cast James's role with an eye toward the religious present. If we go by the numerous discussions that reveal Schleiermacher as the father of modern theology—a hegemonic role that seems fixed to the point of being incontrovertible, even though he has become so connected to Karl Barth that their relationship retroactively colors and even undermines the analysis of his original nineteenth-century position[3]—the fact that William James has not been accorded a similar distinction can only be called striking. Given that Taylor shares a common Catholic sensibility with Hans Joas, could his coming out in favor of James be in part because he likewise wants to create room for the integration of faith commitments, room that he sees our secular age—as did Joas in his critical take on the secular history of religions model—only too eager to deny?[4] James was not Catholic, so while this hypothesis may be plausible in terms of the common convictions and scholarly desiderata for the study of religion held by Taylor and Joas, further analysis will have to show how far their material interest in James goes.

As far as I can see, the interests that bind Schleiermacher and James are, on the one hand, a shared sense of the centrality of human selfhood in religion and, on the other, a shared concern for the delimited integrity of *religious* positions as distinct from moral ones. The point that differentiates both from Emerson, although as we will see not necessarily in the same way, is their respective take on nature as informative, rather than constitutive or formative, vis-à-vis the way religious positions in modernity take shape, both in terms of their anthropological center and their moral tenor.

Although the focus of this chapter is squarely on William James, it also brings us closer, albeit indirectly, to what sets Emerson apart from James's more conventionally modern take on religion and theology. To point out James's approach is not to deny his uniqueness but rather to see it as embryonic for an array of modern approaches from which Emerson can be seen to keep a structural distance. As I indicated, the respective positions of Joas and Taylor—with Taylor displaying at first blush the most material interest in the substance of James's thought—seem partially inspired by their desire to make renewed room for the religious angle in the increasingly secular, post-Enlightenment age. While this chapter has as its primary goal to give

James his due, doing so will allow us at the same time to demonstrate that Emerson is less a thinker on modern religion like Schleiermacher and James, who leave some religious scaffolding intact, than he is a modern thinker on religion, willing to deploy a full array of accordion-like moves to change his religious perspective at will. I come back to the focus implied by this last point in the book's conclusion.

Religion and the Individual Self

An obvious starting point for James's position on religious selfhood, though in other ways perhaps also a somewhat skewed one, is the famous statement found early in *The Varieties of Religious Experience*.[5] I cite it here to continue thinking along the lines set out in the previous chapter on Schleiermacher. What we encountered in our discussion of Schleiermacher was his account of an attempt by the modern religious subject, made to feel destabilized by the Enlightenment erosion of tradition, to reach for a new and more appropriate framing of religion. Although in the worldview of Schleiermacher, and even more in that of his Romantic peers, this framing exhibited cosmic dimensions, it also implied a simultaneous commitment to a form of human self-piloting. In my view this commitment to, and the need for, self-piloting persisted in Schleiermacher despite his increasingly emphatic articulation of the self's foundational dependence on the divine and the concomitant withdrawal from the cosmos it signaled.

This kind of self-piloting sets Schleiermacher fundamentally apart from the thinkers treated in Part 1, including the remarkably self-aware Eriugena, insofar as none of these earlier thinkers ever wanted to move outside the circle of nature. Rather, their response to the overwhelming transcendence of the divine was to enlarge the flexibility of their thinking by creating conceptual niches inside of nature, which allowed them to treat nature piecemeal rather than all at once. When Schleiermacher settles on self-piloting as a way to distinguish himself from the dynamics of nature, we see him instead seeking to preserve—but increasingly also, recalibrate—a connection with the ecclesial tradition that is as robust as it is newly minted.[6]

This grappling of the modern religious subject in search of the appropriate framing of its quest is made even more visible in William James. In clear contrast to Schleiermacher, James's framing is not merely rooted in his concentration on selfhood and simultaneous distancing from religious tradition but appears deeply conditioned by it throughout:

Religion, therefore, as I now ask you arbitrarily to take it, shall mean for us *the feelings, acts, and experiences of individual men in their solitude, so far as they apprehend themselves to stand in relation to whatever they may consider the divine.* Since the relation may be either moral, physical, or ritual, it is evident that out of religion in the sense in which we take it, theologies, philosophies, and ecclesiastical organizations may secondarily grow. In these lectures, however, as I have already said, the immediate personal experiences will amply fill our time, and we shall hardly consider theology or ecclesiasticism at all. (*Varieties*, 34; James's italics)

As is made plain here, James defines modern religion as a set of feelings, acts, and experiences of individual men in their solitude. Theology, which has routinely been seen as the systematic metareflection in which religion ultimately crests, and ecclesiasticism, as the communal but also more hardened institutional and corporate extension of the Christian religion in which theology has traditionally been found embedded, are altogether foreign to what James defines here as religion, which he, with an oddity that doubles as scientific sampling, asks the reader to arbitrarily accept. To underline the phenomenological as well as the contingent, temporal aspect implied in the idea of arbitrariness, as James goes on to assert rather than argue his case, and the sense of random occurrence it connotes, it is important to note that James is at no point interested in speaking about the essence of religion. From the fact that he circumscribes but does not prescriptively impose religion as a set of feelings, acts, and experiences—with which he appears to anticipate the later cultural anthropological and sociological approaches of Clifford Geertz and Robert Bellah as much as he marks future generations of scholars in his field of philosophy of religion[7]—we may reasonably infer that he considers the exclusion of essentialization a productive step forward.

The recoiling from essentialization as a particularly nonmetaphysical (read: pragmatist) way of approaching religion is borne out further by James's relativizing modifier about human beings: "so far as they apprehend themselves to stand in relation to whatever they may consider the divine." Here it becomes clear that relationality, which in the above passage comes across as James's alternative to essentialism, plays an important role in both helping James to keep the dynamics of religion alive along what I consider Emersonian lines and undergirding, in typical Jamesian fashion, the scientific approach to religion. In modernity science has come to play an increasingly important role in the definition and explanation of religion, while also serving as the legitimation for its flourishing.[8] Both its explanation and the legitimation of its flourishing rest for James on having the right scientific paradigm to understand the flow of religion's various relations.[9] Kant's concept of religion within the limits of reason alone appeared to have settled the case of religion in modernity in favor of a scientific approach, science being the systematic extension of rational control. And while Schleiermacher successfully staked out a separate terrain for religion, he did so at the cost of drawing it back within the confines of the ecclesial tradition from which it had just been wrested and, at least in philosophical eyes, liberated. Both Emerson and James would seem to have to reject Schleiermacher's stance as not in alignment with their more emphatically modern takes on religion, which prescind from reliance on past or even present extrinsic structures. But even as Kant's perspective was a soundly rational one, it drew formal boundaries around religion, meaning that he did not address the question of religious experience or of actual religious flourishing directly.

If we zoom in on religious selfhood in James, situating it within the context of relationality, as he prefers, it is clear that the will must play an important role in the further unfolding of his thoughts on the matter. This is clear from his famous essay "The Will to Believe," which was published in a volume bearing the same title that predates *Varieties*.[10] From the rather cynical start of the essay, in which James advocates giving not a justification by faith but rather a justification *of* faith (13), while leaving open the question of the connection between faith and religious experience, he quickly moves on to introduce his readers to the scientific notion of hypothesis, which he explains can be seen as "either live or dead" (14).

Immediately, we see relationality play an important role, for it is not innate properties but rather "relations to the individual thinker . . . measured by [the individual's] willingness to act" that decide whether a hypothesis is dead or alive (14). As if wanting to actualize the hypothesis forthwith, James calls the decision between two hypotheses *an option* and clarifies that what he is after is a so-called *genuine option*, namely of the forced (that is, not avoidable), living (that is, not dead), and momentous (that is, not trivial) kind. With a few handy maneuvers, James stretches the cloth that serves as the canvas of ethics and religion on which he will soon paint his "varieties of religious experience" as a landscape of wildly blooming flowers, the range of their colors determined by his pragmatism, as well as by his moral beliefs alongside his psychological ones.

As James reaches for what he calls a *genuine option*, however, the cloth becomes at times stretched to the breaking point, and anything unhealthy, with which he will be specifically concerned in *Varieties*, and *a fortiori* anything unscientific must be considered compromised—that is, contaminated by traditional, undigested, inherited, and all-around inadequate aspects of religion. To allow a full inspection of what was to become James's *forte*, namely, the understanding of the entanglement of religion and morality,[11] all such elements would have to be cleared aside. But do they really have to be? There appears to be an odd contradiction in James's push for a genuine option in "The Will to Believe" and in the way in which he sees genuine religious experience as most manifest in what he calls in *Varieties* "the divided self." It may be pushing too far to see James considering religion itself as something unhealthy, but there is at least a lingering question in this regard. Related to it, a subquestion emerges as to what exactly is genuine, not only about the will to believe but also about religious experience itself as defining human existence and human selfhood in particular.

"The Will to Believe"—whose title captures the bond of reason and faith rather differently than do other traditional tensive accounts, even if no less indissolubly—is James's attempt not just to wrestle with and get the better of Kant's rational reflex to box religion in, lining its confines with unforgiving and inflexible bounds—that is, the so-called limits of reason. But it is also his attempt to open up religion in a way that refrains from asserting hard dogma, whether religious, as made clear by his comments on Newman, or nonreligious, as evidenced by his gesturing to Clifford's anti-Christian order of the universe (21–22).[12] Interestingly, although we may connect the Jamesian "will to believe" to modern conceptions of selfhood, it appears James would have been wary of the secularization thesis—long before that thesis itself was formulated—with its claim that the ongoing development of modernity would phase out religion. His attunement to temporality would likely have prevented him from settling for any thesis at all.

The upshot of all this is, first of all, that the only certitude that pyrrhonist skepticism leaves standing for James is "the truth that the present phenomenon of consciousness exists" (22). Phenomenologically speaking, this statement may well make James a closer, that is, more straightforward, ally to Augustine in terms of linking belief to consciousness than to Descartes, who is usually credited with being the first modern thinker to echo Augustine's antiskeptical position on selfhood, but whose strong focus on the grounding force of the *cogito* puts him closer to Kant. Second, it means that James is indeed deliberate about the need to be forward looking, resolved as he is to concentrate on the future goal, the so-called *terminus ad quem*, rather than on the origin, or *terminus a quo*, like the more scholastically oriented thinkers of his day (24). The will, then, more central than reason and thus driving James's scientific hypotheses to a considerable extent, lends his thought a directional quality by lifting the self out of whatever rubric or compartment in which it may have previously found itself in the scholastically oriented view. In sum, when we connect the will to the broad process of religious self-piloting, the will is what sets James radically apart from the scholastic developments, as he calls them.

Through his foregrounding of the will, in addition to supplying rational selfhood with added directionality, James also undermines theological scholasticism in another way. While Kant had famously transformed the classical Aristotelian correspondence theory of knowledge, seen as bringing the mind to reality (*adequatio rei et intellectus*), into the new view whereby reality is tailored to the mind's receivership (*adequatio rei ad intellectum*), James's insertion of the will expands the traditional epistemological dialectic to a triadic one. Yet unlike in late medieval Scotism, his focus on the will, because it is on the scale of the individual subject, does not make for a voluntarist climate whereby the divine can impregnate and influence reality at will. While for James reality retains the warrant of a reliable and even amenable cosmos despite an even greater awareness of the hiddenness of noumenal ideality than we find in Schleiermacher, the aegis of a deity is there to galvanize it into the living option that James wants to lay out. Therefore, notwithstanding the moral and pragmatic views for which James is commonly known, we should neither overlook the cosmic allure of these views present in the background nor bypass the role of the will as the vigilant counterforce that awakens the self from cosmic slumber and counters self-resignation.

Crucially, it appears that James's chosen title, "The Will to Believe," has itself also a kind of relationality to it, not unlike "The Sentiment of Rationality,"[13] the earliest essay in the same volume and one we will discuss shortly. In exploring the relation between will and belief, it is useful to dwell on James's passing comments on Pascal's wager. James adduces the famous

wager to test whether one can steer one's volition to belief and thereby for-
tify the move from self to God. It is tempting to infer that we see James the
scientist at work here, as he claims a little later that "in our dealings with
objective nature we are obviously recorders, not makers of the truth" (26).
In ascertaining the validity of the wager, James seems to want to tease out
the primary value of any such move as being the product of a scientific project. In line with this he makes the point that a Muslim would likely not be
persuaded when the Mahdi promulgated the Pascalian argument. His suggestion here seems to be that the wager's validity pertains to Christianity
alone and that this religious particularity threatens to make it unscientific.
While it appears as if James the scientist is dissecting Pascal, inferring that
his apologetic claim loses its persuasive power when deployed outside the
religion from which it sprang, a comment that follows signals otherwise.
In a throwaway line James states that Pascal's wager is only a living option
for those who have a preexisting tendency to believe in masses and holy
water (16), by which he references those believers who are embedded in a
sacramental approach. Here he betrays a kind of Protestant repugnance at
what he apparently considers a Catholic argument—namely, that religion
is institutionally dependent rather than originating in the individual. We
can find similar anti-Catholic moves in *Varieties*, but the bluntness of his
comment here makes it rather stand out, as it is simultaneously very much
on point and completely off-kilter. Insofar as there is indeed a Catholic
susceptibility, though with Augustinian overtones,[14] that animates Pascal's
notion of religious faith, James may be on target, even if "holy water and
masses" is an irresponsibly reductive shorthand term for the sacramental system.[15] But he is off-kilter insofar as the aspect of sacramental attractiveness
is peripheral, if not completely forensic to the larger question of whether
Pascal's wager works or not, and hence should have no impact on James's
theme of the will to believe.[16]

   While we often find James advancing the more scientific posture of a
recorder of truth in *Varieties*, in Lecture 19 he discusses the question of a
Protestant and Catholic aesthetics. He formulates the difference between
them as follows:

> The strength of these aesthetic sentiments makes it rigorously impossible,
> it seems to me, that Protestantism, however superior in spiritual profun-
> dity it may be to Catholicism, should at the present day succeed in making
> many converts from the mere venerable ecclesiasticism. The latter offers a

so much richer pasturage and shade to the fancy, has so many cells with so many different kinds of honey, is so indulgent in its multiform appeals to human nature, that Protestantism will always show to Catholic eyes the alms-house physiognomy. The bitter negativity of it is to the Catholic mind incomprehensible. To intellectual Catholics many of the antiquated beliefs and practices to which the Church gives countenance are, if taken literally, as childish in the pleasing sense of "childlike"—innocent and amiable, and worthy to be smiled on in consideration of the undeveloped condition of the dear people's intellects. To the Protestant, on the contrary, they are childish in the sense of being idiotic falsehoods. He must stamp out their delicate and lovable redundancy, leaving the Catholic to shudder at his literalness. He appears to the latter as morose as if he were some hard-eyed, numb, monotonous kind of reptile. The two will never understand each other—their centers of emotional energy are too different. (*Varieties*, 363–64)

So much for a Kantian religion within the bounds of reason as a viable container, given James's commitments. In James's view such bounds simply cannot keep experience in check, and the powerful ecclesiasticism that has taken flight in Catholicism finds itself victorious over diagnostic reason as soon as one unleashes its aesthetics. James's awareness of the limitation of rational cogency proves a powerful tool here. Whereas in his earlier dismissal of Pascal's wager he may have inflicted blunt trauma but no effective wound, here he seems ready to turn the tables and crush his own Protestant tradition. With the balance now set straight, we can better see that underlying his confessional descriptors both here and earlier, there is indeed a budding attempt at scientific typology. His unscientific designations thereby emerge as the products of an observing and curious recording that dares to tread outside the box of ordinary confessional description: the almshouse physiognomy, the monotonous reptile.

It is hard to suppress a sense of recognition when reading these aesthetic musings, especially since they too are at once spot-on and deeply caricatural. Clearly, this is one strong aspect of James's decision to bring in the world of experience, of aesthetics, of sensory attraction, which allows him to make us alternatively smell and be revulsed or hear and be edified. But whatever our reaction may be, it cannot fail to make plain religion's phenomenal impact. The immediacy of religion in James gives it a nonintellectualist quality. It does not have to be funneled through dependence on the infinite to be legible but can at once be widespread and fine-tuned. *Varieties* spins for us a fine gossamer of religious experiential threads that are in the end as personal as they are irreducible but do not rule out that the recorder of these observations—that is, James himself—is at the same time much more than a mere recorder. By putting his own spin on things, he adds a creative twist to what it means to record truth; as we will see, that spin is also part of the religious data that is assembled. As David Lamberth puts it in *William James and the Metaphysics of Experience*: "Reflection (or conceptual thinking) for James is fundamentally an additive process, a process that contributes to reality, building it out by the edges. . . . Interestingly, the products of reflection, or for that matter philosophy, for James are not fundamentally separate from the realm out of which they are abstracted and to which they add."[17] In other words, James studies religious experience, even if he can also undergo it, which again confirms his preference for a functional and pragmatic approach rather than a foundational or grounding one. Like Emerson, James goes against the tide of the times, which, in Schleiermacher, Nietzsche, Newman, but also Darwin, who seems to have especially weighed on James's mind, were primarily focused on the probative value of history and genealogy.

Whatever typology James eventually settles for, it does not impose a one-size-fits-all rubric of human selfhood but is better seen as his attempt to register seismographically any and all human tremblings of a psychological, religious, and moral nature, and account for them in their "variety": an open phenomenological, rather than specified scientific, term. He appears to approach religious experience from the perspective of a self that can alternately be empathetic, compassionate, scientifically engaged, distant, or curious but that generally is not solipsistic,[18] as that would disallow the human relation to "whatever they may consider the divine" to come through.

Yet where does this leave us vis-à-vis Pascal's wager? Does James's dismissal not imply it to be so emically constructed that it must ultimately fail to be persuasive? And is the reason for its failure perhaps that it borders on the solipsism James rejects, leading him to conjure up the holy water and the masses in an effort to make Pascal's failure resonate against a reality that is recognizably material and sensual? But while his repudiation of Pascal may point to the kind of rejection that he would also have to apply to many philosophical arguments more broadly—namely, that their analytic quality makes them incapable of piercing through the synthetic web that is reality—such a rejection gestures to a reading of Pascal that is ultimately more Cartesian or stand-alone than Augustinian or grace-filled. Precisely his focus on the conscious rather than the rational self should have permitted James to extend the desire for the recording of truth into the space of interiority, even if phenomenologically construed, in a way that banishes the refuge of solipsism and allows entry into the dimensionality of a new Emersonian circle—a circle that allows persons to unfold themselves not only outside and forward but also inside and onward, into a new identity of religious believers. Indeed, in the Catholic tradition, even of the Jansenist brand that fascinated Pascal, there remains a correspondence of the interior self not just with the ecclesiasticism that James flags as a mark of superstition rather than faith but also with the agony and agency of divine grace. In my view, then, one can thus read Pascal in such a way that the wager allows one to prop up one's ecclesiasticism from the inside out, since there exists a subtle link between interiority and ecclesiasticism rather than forcing one to dispense with it.

Faith and the Moral Universe

Setting the Emersonian circular, multidimensional self aside for a moment, I propose that the question about the Jamesian self is to what extent it fits into—and hence is regulated, if not determined by—a larger cosmic whole. Addressing precisely this question, the philosopher of religion Wayne Proudfoot has called attention to the notion of an unseen order in James. He takes as his starting point James's question, formulated in an article that later became incorporated in his essay "The Sentiment of Rationality" in the collection *The Will to Believe*, of whether we live in a moral or unmoral universe. After quoting Proudfoot on James's position, I will discuss James's views in "The Sentiment of Rationality" in more detail.

At the end of his article "William James on an Unseen Order," Proudfoot writes:

> The world of nature as we perceive and experience it is shaped by our categories, including moral ones. All experience is interpreted. Histories of representations of nature in literature and the arts as well as the sciences are social and cultural histories. But it is important to recognize this and not to assume that our moral orderings are grounded in nature or metaphysics in some way that is prior to human thought and activity. When we recognize these orderings as interpretations, we can be free, with the romantic

poets, to personify and project, and to engage imaginatively with the natural world. This engagement can become an avenue for self-knowledge.[19]

With his comment on, but also criticism of, James's unseen order, Proudfoot seems to want to keep open the option for humanity that, in principle, this order is one that must be interpreted and can to that extent also be reshaped.[20] Pointing out the difficulty involved in such a reshaping leads him to make useful connections between James and Freud, Marx, and Nietzsche. Proudfoot further elaborates his position in the essay "Pragmatism and 'An Unseen Order' in *Varieties*," in which he quotes James's view of the unseen order that opens Lecture 3 from *Varieties*, entitled "The Reality of the Unseen." Here James writes, "Were one asked to characterize the life of religion in the broadest and most general terms possible, one might say that it consists of the belief that there is an unseen order, and that our supreme good lies in harmoniously adjusting ourselves thereto. This belief and this adjustment are the religious attitude in the soul."[21]

Quite different from the earlier definition of religion from *Varieties*, here the relational element in James comes fully to the fore, framed as the belief in an unseen order. He gives it the added qualification "that our supreme good lies in harmoniously adjusting ourselves thereto," as if bringing it all back once again to man in his individual—albeit cosmically mitigated—solitude, and states that "this belief [in the unseen order] and this adjustment are the religious attitude in the soul." In sum, belief in the unseen order is crucial for James: it is characteristic of religion and is nested in the soul or, rather, constitutes the soul's religious posture.

But what about Proudfoot's initial critique, elaborated more pointedly in his later article, that James lacks adequate attention to historical context and thereby overlooks the possibility that the unseen order can be reshaped?[22] For Proudfoot this is a grave issue. To the extent that he wants to further James's move from theology toward a science of religions, he pointedly disagrees with James "on the need to 'remove historic incrustations' from dogma and worship and, instead, confront the spontaneous religious constructions with the results of natural science."[23] For Proudfoot it is simply not possible to develop a science of religions without a proper understanding of the historical circumstances of the sources that are consulted, which he considers a *sine qua non* for the effective separation of the essential from the individual that James desires. Proudfoot's concern is echoed by Richard Rorty vis-à-vis the seeming closeness—for lack of a systemic articulation on James's part—of his view with naturalists like Marx, Nietzsche, Dewey, and Freud. Rorty thinks that, had James carried on his pragmatism, *religious* could have become a synonym for "vitally important to a person's self-image," bringing him close to other thinkers like Nietzsche.[24] Had James developed greater historical insight, per Proudfoot's preference, it might well have allayed James's own concerns about naturalism through the attenuating influence of historical origins and context. Proudfoot's summary diagnosis is that James's fear of naturalism leads him to make insufficient use of *Geist*, as a result of which he draws insufficiently on the *Geisteswissenschaften*. Let us see if his diagnosis is correct. Further contextualization of the historical figures on whom James draws, which Proudfoot considers indispensable for an adequate science of religions, would have made James aware of what Proudfoot calls "the multiple moral resources that are available to people."[25] Had James brought the unseen order into our line of sight, that order would have become more susceptible to human action and amenable to human reshaping.

To get a firmer grip on James's problems with the morality of the universe, with religion as the belief in the unseen order, and with naturalism and the rationality underlying it, let us engage James's full essay, "The Sentiment of Rationality."[26] It opens as an essay on philosophy, not religion, as James states his goal is to "attain a conception of the frame of things which shall on the whole be more rational than that somewhat chaotic view which everyone by nature carries about with him under his hat" (57). It is not the framing of things but the underlying or enveloping *conception* of the frame of things that ought to be rational, which philosophers recognize through resulting attributes like peace of mind and rest, not unlike what he in *Varieties* will see as the result of conversion. James defines rationality as an anesthetic state, which reeks of the opposite of Stoicism—namely, the feeling of the sufficiency and absoluteness of the present moment. This feeling is what James calls the "sentiment of rationality": "Whatever modes of conceiving the cosmos [that] facilitate this fluency, produce the sentiment of rationality" (58). The analogy with conversion extends even further, in that "being vouches for itself and needs no further philosophic formulation" (58).

James goes on to research how such fluency may be obtained. The theoretical way aims at reducing the unwieldiness of original data to a labor-saving simplicity, therein echoing the parsimony that is philosophy's passion (58). The knowledge of things in their causes does not bring relief; only universality can do so. James sees the passion for distinguishing as the sister to the passion for simplicity, the former resting on familiarity and acquaintance with concrete particulars (59). While Hume and Spinoza serve respectively as the discrete models for these divergent passions, James considers a compromise between them to be the only possible philosophy. Yet while heterogeneous things can be subsumed under kinds, and the relations between them classified as laws, the fullness of the truth is not known by them (61), because this kind of "progress" does not extend beyond the teleology of classification, abridging rather than manifesting life. In the case of an exhaustive classification, however, should we deem the universal principle in which it rests a rational one? (62). Judging in the negative, and deeming nonentity to be the parent of the philosophic craving, James states that "absolute existence is absolute mystery" (63). Skepticism appears to rear its head here, a problem familiar from Emerson, which James wraps in with the new claim that "a possible other than the actual" (64) may still haunt us. James states that to accept God's fiat in physics and morals serves in fact as an ultimate datum for ordinary men and allows them to put their doubt to definitive rest, even while for others that role is fulfilled by the embrace of brute fact (65).

Logical tranquility not being enough, ecstasy provides us with another take on mediation. In this vein James invokes Wordsworth, who said, "thought is not; in enjoyment it expires" (65). For James, the benefit the religious person has over the philosopher is that he accepts, in his heart, the world as rapturously complete. But while philosophy cannot rise above empiricism, which is itself incapable of capturing the ultimate wonder that only some will fully comprehend, the method of the heart that mystics follow suffers from a lack of universality, according to James.

Switching next from theoretical to practical rationality, James is keen to have a "definition of the world which will give back to the mind the free motion which has been blocked in the purely contemplative path [that may] make the world seem rational again" (66), adding that what awakens the active intellect will be the better choice over what awakens only the heart. But which tests of rationality would be suitable "for our aesthetic and practical nature"? Given the relational quality of rationality, James makes a crucial move when he foregrounds the relation of a thing to its future consequences. The "ingredient of expectancy" (67) is never far from our consciousness, according to James, and the philosophic datum one embraces should therefore banish uncertainty from the future. Next to "substance" in a Kantian, noumenal sense, "immortality" has served this purpose in religion, even as it has also provoked philosophical ire against it, kindling the skepticism of empiricists (70). While the philosophical craving to define expectancy must be met, James qualifies its success on the level of the universal datum, not by determining expectancy either in optimistic or pessimistic terms but by stating pragmatically that "it must define the future *congruously with our spontaneous powers*" (70). This accords with his audacious rewiring of the historical-philosophical quest to align with his view that the intellect is built up of practical interests, marked by the direct question: "*Was fang ich an?*" (What is to be done?) (72). Rather than creating contrasts, however, James holds that *comprehension* and *acquaintance* go hand in hand, just as *wissen* (to know) and *kennen* (to know intimately) go hand in hand, thereby instructing us how to behave toward a thing (72), as anesthesia for him is simply not permitted (73).

Zooming in more and more on religion, James is not averse to engaging history to bolster his claims, even as he exorcises skepticism. Thus, he holds that all great periods of revival have encouraged humanity to think that reality's inmost nature is compatible with human *powers* (73). Mentioning early Christian repentance as an appeal straight to the heart of God, yet skipping the Middle Ages entirely, he sees the *sursum corda* of the Platonizing renaissance as an affirmation that the archetype of verity lies in the things themselves, and he commends Luther and Wesley for their joint mobilization of faith and self-despair, which seems to betray how they incorporate faith's own counterfactual within it (74). Thereafter, philosophical history fans out to Rousseau's harmony with nature; to Kant, Fichte, Schiller, and Goethe for their exhortation to use all of one's powers; and to Carlyle's valuation of work, only to crest in what he considers to be "Emerson's creed that everything that ever was or will be is here in the enveloping now," which he encapsulates by his quotation of Emerson, "He who will rest in what he is, is a part of destiny" (74).[27] Clearly, the direct appeal to man—like Emerson, James addresses humanity as the *Son of Man*—to stand on his own two feet is the one revelation that holds (74), even while there is room for temperamental differences; such differences explain why idealism, centered on intimacy and atonement, appeals to some, and materialism, centered on agnosticism and brute fact, to others (75).

Be that as it may, what James holds Christianity to contribute and philosophy to overlook is the element of faith, defined as "the readiness to act in a cause the prosperous issue of which is not certified to us in advance" (76). Morally, such readiness is not unlike courage, but in philosophy it seems only permitted to allow for the uniformity of nature's course (76). Echoing Pascal, but on a universal rather than a personal scale now, James calls faith a working hypothesis that may defy ages (79), referencing his engagement of history. Faith is thus truly a universal. With respect to the meaning of God, immortality, absolute morality, and free will, there may be minor creedal differences, but Christians generally uphold their truth until the day of judgment, with their "volume of . . . feeling" compensating for whatever lack of positive arguments they may have (80).

Not only do we act out of faith, according to James, but there is a certain class of truths for which faith is not only pertinent but indispensable, insofar as these truths cannot be valid until and unless faith makes them true, since faith creates its own verification (80). Whereas some truths are given irrespective of human desires, many more depend on subjective energy, which should therefore be implicated in any universal philosophy. If M is the universe without the thinker's reaction on it and $M + x$ includes the reaction and its results, given that x varies, it will always affect *the whole in which it is embedded*, according to James (81). Even evolutionists, he argues, have to factor in that the course of destiny may be altered by individuals. "Success depends on energy of act, with energy depending on faith that we shall not fail, which in turn depends on faith that we are right, which faith verifies itself" (83). The same applies to the question of our pessimistic or optimistic take on the universe. The highest good depends on whether or not we get "our proper life," which means that our faith bears in us a moral energy that aids us in persevering toward our goal (84).

All of this allows James to now restate the question of philosophy as whether we live in a moral or unmoral universe, the point that Proudfoot highlighted. Warding off materialism, James brings in the matter of faith and asks whether it is objectively "best for the cosmos to have this me" (85) or whether there is a more subjective kind of moralism at play. With the verification of one's belief lying in the posture assumed by the individual, the objective or absolute moralist will hold that the universe is moral and have his own morality to prove it, while the subjective moralist refuses tragedy and wraps himself in the anesthesia of the moral skeptic (87). James famously concludes by seeing nature as furnishing us with two keys with which to test the lock: if we try the moral key and it fits, it is a moral lock; if we try the unmoral key and it fits, it is an unmoral lock (88). The "co-operation of generations is clearly needed to educe it," for such decisions are not individual but cross-generational, a kind of human solidarity that James considers "a patent fact" (88).

For James, then, no philosophy will ever be universally rational that does not establish expectancy to some extent, thereby appealing to the higher powers of our nature. Faith, as one of these higher powers, must play an important role in philosophy, the more so since faith brings about its own verification. Faith also offers an escape into another, nonpedantic realm that is located above the universal doctrines to which we subscribe (89).

The Quest for Religious Naturalism 1:
A Divided Self and a Two-Storied Universe

From the above, as well as from all we know James to generally stand for, it is obvious that his quest is not to establish a natural theology, not even of the psychological kind; nevertheless, it seems fair to describe and analyze his scientific pursuit as motivated by an interest in religious naturalism. By that I mean two things: (1) religious experience, not unlike faith in "The Sentiment of Rationality," makes a material difference in the universe inasmuch as it calls human beings to actions that are religiously motivated; and (2) religion is therefore to be treated as a part of human beings' natural outfit, from which it can be inferred that any philosophy of human nature that disqualifies the importance of faith is to that extent deficient.

For all my references to selfhood, it is the notion of a divided self, treated in Lecture 8 of *Varieties*, that has made James famous. This puts another interesting spin on his engagement with the Christian tradition, inasmuch as James mostly abstains from the Pauline language of the divided will that was to have such a long afterlife from Augustine's *Confessions* onward. If we connect Pascal's wager as discussed in "The Will to Believe" with the language James introduces here, we might consider recasting Pascal's purpose as an attempt to overcome, or perhaps circumvent, the notion of a divided will by synchronizing one's internal will with one's external religious life.[28] Such a reading would be a way of broadening its applicability beyond the Christian boundaries to which James objected in his essay "The Will to Believe." But James's goal is neither to justify Pascal nor to cope with the divided will by following in his footsteps but rather to diagnose and analyze what causes division in human nature. Ann Taves has pointed usefully to what she calls the fissiparous nature of the Jamesian self, tracing its origins to research with which James had recently become acquainted by Pierre Janet (rather than the oft-assumed Frederick Myers), which he put in service of his greater "task to formulate a hypothesis about religion to which physical science need not object."[29] In Lecture 8 James makes occasional use of the language of will, but the larger scope within which he treats it is that of selfhood. He thereby gives us yet another signal that, his pragmatism notwithstanding, the religious dial had shifted for him from voluntarism to questions of selfhood, consciousness, and existence.

Predictably, the way in which James thinks the division of the self ought to be overcome is through the attainment of unity, which would seem to involve some collaboration of the human will. Unity can be achieved either by means of a gradual process or by sudden conversion—what we might call a *coup de grâce*. The choice between these alternatives does not necessarily pit the self against the divine, as transcendence can rule the gradual process just as much as it can provoke a more radical, sudden conversion. So, it is not so much human selfhood and divine transcendence that are the psychological goalposts regulating James's view of religious development, as it is the more poignant pair of solipsism and grace, the latter reflecting the way in which divine transcendence, when pragmatically tilted, can produce healing and, therein, unity. More surprisingly perhaps, James does not give us any assurances, nor does he stipulate that the unity that is to be attained is necessarily religious.[30]

Here, then, we can detect in earnest the contours of James's thoroughly modern treatment of religion. It differs from Schleiermacher's exactly a century before in the sense that, as we have seen, James considers religion to be an important part of our anthropological makeup, but he does not thereby let religion define human nature entirely, or even foundationally, as if humanity was originally religious before it was secular. For good or for ill James disagrees with Schleiermacher, Newman, Nietzsche, and Darwin, but he is in full agreement with Emerson, in that he does not assign history probative value but rather uses it to prop up the innate force of the human beings whom he observes. Yet insofar as James is engaged in a project that, at the time of *Varieties*, appears to prioritize "thinking human nature" over "thinking nature," religion holds a steadfastly powerful appeal to him, even if he realizes at the same time that the religious person cannot stand in for all persons. To the extent that the universal ought to have a touch of the distinctive, it is religious faith that makes it so.

Thus Charles Taylor has voiced an interesting position in singling out William James as a possible new beginning for the study, or philosophy, of religion in our own era. This is my somewhat overstated interpretation of his point that James still has something to say to us today.[31] To see James inaugurate a new era for the study of religion seems indeed a fitting role for him, one that he has long been denied, to which the multiple voices in the centenary volume on *Varieties* attest. Despite Schleiermacher's iconic reputation as the father of modern theology, he moves within a narrower circle insofar as he does not engage nonbelievers or those whom Taylor insightfully calls "on the cusp."[32] Instead, Schleiermacher focuses his dialogue on the concrete group of cultured despisers with whom he felt enough affinity—and no less competition, it would appear—as to want to bring them back into the fold of the church. Clearly, James's reach is much wider, and at the risk of making his project seem altogether grandiose, his scientific posture paired with his freedom from any overt theological commitments makes him remarkably unfazed.

If we can agree on this observation, it would mean that Taylor, whose Catholicism I referenced alongside that of Joas, may actually be on a quest that, though largely in agreement with the scholarly voices in the centenary volume, is the de facto mirror image of Joas's project. For, while Joas expressed interest in developing an approach to the study of religion that allowed for religious commitment, in his case a Christian one, Taylor's project, even if he does not fully elaborate it, is to have a philosophy of religion, very closely positioned to a theology it would seem, that is decidedly modern insofar as it recognizes the validity of unbelief and the secular. Both thinkers are justified in turning to James.

James's stance of recognizing religion as constitutive of our anthropological makeup, even if not exhaustively so, cannot fail to raise the question of relativism. For, to what degree does his project of "thinking human nature," with faith and religious experience as key aspects within it, influence the nature of his thinking about human nature and the universe as its enveloping framework more broadly? At this point in the discussion I want to return to solipsism and grace as the two goalposts inside of which James's debate on religious experience is conducted. I see solipsism as located on one end of the spectrum since it rules out all connection with other human beings and must therefore thwart any scientific approach that would involve some degree of intersubjective contact. Grace, positioned on the other end of the spectrum, appears to suffer from a similar disadvantage insofar as the overpowering force of divine transcendence removes independent human access to it and clouds the transparency of any possible divine-human interaction. Grace as undiluted divine transcendence risks obliterating if not entirely canceling out the factor of human experience and accountability, while solipsism does not just reject divine interference but would also seem to block any and all traceable impressions of it. Of course, we still have to determine whether or not James invokes these same goalposts as his categories of choice.

If we look first at Lecture 8, "The Divided Self and the Process of Its Unification," we see how James maps out the divided self in anticipation and preparation for the following two lectures on conversion (Lectures 9 and 10).[33] As far as famous conversions go, often provoked and preceded by a jarring sense of division, the case of Augustine receives exceptional praise, with James lauding it as the best description ever. Still, the focus—and the lasting source of James's fascination, one may surmise—is squarely on the cases of Tolstoy and Bunyan, both of whom are situated in the modern era and are therefore fitting protagonists of James's deeply modern, scientific agenda in *Varieties*.

Taking a scientific posture but aiming at a diagnosis that best accommodates human religious experience in its acclaimed variety, James admits to what he calls two idealized abstractions to regulate his observations. There is the once-born, healthy-minded soul whose life in the world is a one-storied affair, and who is hence put on the side of pure naturalism, while placed on the other side is pure salvationism, as he calls it, where we find the sick soul who must be twice-born in order to be happy. As far as religious experience goes, the world is apparently a double-storied mystery for James. Not solipsism and grace, then, but naturalism and salvationism are the two goalposts that James selects, the former reflecting a one-storied universe, the latter a two-storied universe, as the divided self can only find redemption when born again. Betraying his sense of compromise, James is quick to add that "the concrete human beings whom we oftenest meet are intermediate varieties and mixtures" (140).

Given the lecture's title, "The Divided Self," it is no surprise that he defines the sick soul in terms of a discordancy or heterogeneity, "an incompletely unified moral and intellectual constitution" (140). Clearly, religion bleeds into morality and vice versa, although James's approach is to proceed through a focus on *character* as the site where religion and morality meet or blend into one another. The invocation of character allows James to have an approach that not only resonates with a Freud-like *insight* into the subconscious but that on a somewhat higher, aggregate level affords him a broader, more ecumenical grip on how, in the order-making process, unhappiness yields religious melancholy and "conviction of sin." Under this rubric he ranks not only Protestant Christianity but also the likes of Victor Hugo's Mohammed, St. Paul, and, obviously, Augustine. Remarkably, the Protestant aspect is here not characterized through forced isolation but mined instead for usefully descriptive, as well as integrative, indicators of a condition found among adherents of various faiths. It should again be clarified that, for James, the unifying process that the divided self must undergo to arrive at happiness does not have to take a religious form (146). Religious regeneration is one species in a broader genus, which also includes patriotic devotion, for example. One wonders whether sports or celebrity idolization today might not also count. While the transition to such regeneration is often marked by conversion, indicative of a religious resolution, it can conversely also be captured by counterconversion, that is, a falling away from religion rather than a birth into it.

A surplus of religious insight and mysticism procures relief for the divided selves, cresting in their profoundest happiness, while it brings along unity and peace (*pax et concordia*) for these torn souls. Overall, it appears James gravitates more toward the gradual process of what he calls, with a medical term, *lysis* (indicating gradual deterioration) rather than the sudden process of *krisis* (indicating a sudden rupture), choosing Tolstoy and Bunyan as his examples (153). James ascribes Tolstoy's conversion to his rejection of the elite intellectual culture to which he belonged but in which he did not feel at home, and which over time he came to replace with the simplicity of "life." James quotes Tolstoy: "To acknowledge God and to live are one and the same thing. God is what life is. Well, then! live, seek God, and there will be no life without him" (154). For James, Tolstoy's melancholy can only truly be overcome when the clash between his inner and outer life is resolved. In Bunyan the same melancholy is provoked by scriptural texts that alternately give him comfort or disquiet him, until relief is brought at last by the belief in Christ's salvation, prompting him to an active ministry that is fueled by the knowledge that what Bunyan called "the tempest," including his twelve-year prison sentence for nonconformity, was almost over.

As James draws his lecture on the divided self to a close, his conclusion is that Tolstoy and Bunyan never became completely healthy-minded, "as their redemption is into a universe two-stories deep" (155). Projecting the dynamics of their salvation onto a wider screen, he leaves unclear what he means by a two-storied universe, but there is a sense that faith has brought them an escape beyond the universe as it is encountered in the reality of their lives. Yet there is a further moral to the story as well, a religious one according to James, that should not be overlooked. James states that "they did find *something* welling up in the inner reaches of their consciousness, by which such sadness could be overcome" (156), which seems an oblique reference simultaneously to his religious realism and to his pragmatic take on it. This inner faith, called a "something" and therefore indicating substance even though it falls under the category of experience, sustained Tolstoy in defiance of any outward institutional values. James sees this defiance resembling the attitude of Bunyan, who also "leaves this world to the enemy" (156). Both Tolstoy and Bunyan thus operate on the existence of a difference between the inner self and the outer world. But it is the reified inner self as an awakened force or a sustaining grace—to use that overt religious term here not to indicate the opposite of solipsism, as discussed above, but as a dynamic experiential marker—that is their compass in the world, both morally and religiously, and so, one assumes, drives their actions in the cosmic order of things.

James continues his narrative in Lectures 9 and 10, both entitled "Conversion," by further scrutinizing and theorizing the problem of the divided self. Although in these next two lectures James is tackling the same topic of conversion, the opening of each shows him to take rather different positions. At the opening of Lecture 9 he is set on analyzing the process of unification that he already brought under way in Lecture 8. Leaving behind the question of whether it is sudden or gradual, he focuses now on regeneration and, therewith, on unification and the healed self's being consciously right, superior, and happy. James describes the effect of conversion: "to say that a man is 'converted' means, in these terms, that religious ideas, previously peripheral in his consciousness, now take a central place, and that religious aims form the habitual centre of his energy" (*Varieties*, 162).

Before James arrived here, he discussed the notion of transformation and spoke of emotional "alterations and alternations" that seem to underlie the divided self (161), all the while allowing the self to harbor different groups of aims in simultaneous coexistence. He dispelled the idea that the real self is ruled by *velleitates* (whimsies) (161), even though he spoke about personal desire and volition as leading to the alteration of emotional excitement. James seems to have been after the demarcation of a spatial field. Replacing the traditional ontological terminology of mind or soul, his talk about alterations and alternations creates room for what he calls the habitual center of one's personal energy. Conversion in this scenario succeeds when religious aims can somehow make their home there.

Oddly, Lecture 10 opens far more traditionally, with a focus on instances of instantaneous conversion, especially of St. Paul, in which lightning bolts radically divide the old and the new life "in the twinkling of an eye" (178). James again makes a direct connection with Protestantism, stating, "Conversion of this type is an important phase of religious experience, owing to the part which it has played in Protestant theology, and it behooves us to study it conscientiously on that account" (178). To the extent that his foregrounding of Protestantism suggests a bias, it would appear to undermine James's scientific aspirations. If we suspend judgment for the time being, the best we can say is that a functional connection with Protestantism helps him to illuminate the meaning of radical conversion.

If we go back to the first lecture on conversion, Lecture 9, we find James asking *how* the excitement in a man's mental system shifts, and *why* the aims that were peripheral before become central now (162). Psychology does not appear to give him an answer beyond what he calls "the hackneyed symbolism of a mechanical equilibrium" (163). Going through various possibilities of conversion—as similar to adolescent spiritual maturing or to the regeneration and sense of rescue felt by alcoholics—James comes to distinguish between conscious and voluntary and unconscious and involuntary ways of converting. With the volitional type of conversion the healing is gradual. But James moves on quickly to zoom in on the so-called *type by self-surrender* (169–70), which stands out for him because of its remarkable subconscious effects. Anticipating the opening of the next lecture, James adds the comment that ultimately even gradual conversions have moments of self-surrender, thus wrapping the durable in with the instantaneous in a counterintuitive move that can only really be explained if the instantaneous is accorded the higher prerogative. The ultimate necessity of self-surrender, as James sees it, arises in relation to what he calls a negative quality, as there comes a moment when the incompleteness that characterizes "sin" can no longer be escaped and needs to be overcome. This negative quality is mirrored by the positive quality of striving toward righteousness, which he explains in words he has derived from Starbuck.[34] James then states in his own words: "When the new centre of personal energy has been subconsciously incubated so long as to be just ready to open into flower, 'hands off' is the only word for us, it must burst forth unaided" (*Varieties*, 173).

It is worth mentioning as an aside that Emerson, though not treated as an explicit source here, plays a role in the background, as if he is hardwired into James's thought and only on occasion brought out explicitly as an oracle. When, earlier, James gave the example of how the mind can be *jammed* before it relaxes in conversion (169, 172), he compared it to how a name fails to come to mind unless one gives up trying to think of it, at which point it appears "as casually as if it had never been invited" (169).[35] A trace of Emersonian *casualness* may thus well linger in James's preference for a hands-off approach.

Further betraying his Protestant sensibility or, perhaps better, congenitally unable to let go off it, James considers the crisis of self-surrender to be "the vital turning point of religious life, so far as religious life is spiritual and no affair of outer works and ritual and sacraments" (173). A likeminded take on the history of Christianity follows, in which he moves from Catholicism to Lutheranism, on to Calvinism and Wesleyanism, and even to pure liberalism and transcendental idealism, both of which are placed outside the bounds of what James calls technical Christianity. Remarkably, we find ourselves back at the opening of *Varieties* here, with its idea of "religion as *the feelings, acts, and experiences of individual men in their solitude, so far as they apprehend themselves to stand in relation to whatever they may consider the divine.*" But this original notion has been concretized now into "the idea of immediate spiritual help, experienced by the individual in his forlornness and standing in no essential need of doctrinal apparatus or propitiatory machinery" (173). The individual here is less theoretically sophisticated but more emotionally urgent, marked by forlornness and a tinge of despondency, if not despair.

Aware of the discrepancy between psychology and Christian theology, James wants to study them side by side, as he ultimately does not want to rule that everything in the subconscious also originates there, leaving room for the supernatural operations of the divine to enter in. In fact, he sees the divergence of psychology and theology as temporal rather than causal: the exhaustion of the struggle that is succeeded by the entrance of a higher emotion (176).[36] Still, the cases of instantaneous conversion, tied as they are to the impact of divine grace, keep James under their spell, and it is to them that he turns in Lecture 10.

After commenting on the importance of conversion for Protestant theology as justification for his treatment of it, it is notable that when giving his first case—opting to give cases before generalizing—James concludes it by ranking the example of Mr. Alline with those of Bunyan and Tolstoy, mentioning that his "redemption was into another universe than this mere natural world, and life remained for him a sad and patient trial" (180). Rather ascetically, if not necessarily catholically, this "other world" is marked by the absence of carnal pleasures. Yet James, having first cited the case of the French Jew Ratisbonne, who converted to Catholicism, a case he later discusses at greater length, does not pursue this Protestant tack. Taking a different approach, he criticizes mainstream Christian denominations altogether as depreciating instantaneous conversion in favor of their institutional demands, which, for James, "are practically supposed to suffice to his salvation, even though no acute crisis of self-despair and surrender followed by relief should be experienced" (186). The only exception is Methodism, which James regards as following the profounder spiritual instinct and which, unsurprisingly in this context, he sees as a less hierarchically conceived, American, and antithetical version of its progenitor, Anglicanism.[37]

James's next big question in Lecture 10, as he has not yet made a definitive choice between the psychological or the theological, deals with the problem of grace as the supernatural aspect hovering over conversion in another way. He asks whether conversions, which he sees as "one of [human beings'] most curious peculiarities," are to be ranked as miracles (188). This brings him back to the psychological terminology of the field of consciousness that he has used before: the shifting centers of personal energy and the incubation time needed for subconscious motives to ripen. Using the language of a spatial field allows James to invoke the indeterminate nature of margins (189). In the ensuing debate he states that a magnetic field surrounds our center of energy, for which the compass-needle indicates the present state of consciousness.

With this move toward spatialization, James shifts religious debate from a set of temporal pendulum swings between solipsism (human) and grace (divine) to a rolled-out, horizontal landscape of center and margins, even if the vertical dynamics of the two-storied universe in "The Sentiment of Rationality" are not entirely left behind in the use of the conscious and the subconscious. The move toward spatialization allows him to widen the scope and refocalize by bringing in the extramarginal treasure trove of memories, thoughts, and feelings that can be conjured up through what James calls "signs" and, thus, brought into play. It is Myers's famous research on automatism, in which he used hypnotic suggestion to condition certain "uprushes" into one's ordinary consciousness, which has laid bare for James the role of "signs" (191). But James's interest prompts him to move from the eccentric back to our natural constitution and prods him to ask whether there is not a source for automatism in the subconscious. Going back to the cases of instantaneous conversion, which remain at the center of his attention in this lecture, James sees them as presenting a simple psychological peculiarity that contains "a large region in which mental work can go on subliminally, and from which invasive experiences, abruptly upsetting the equilibrium of the primary consciousness, may come" (193). For James, the Methodists, whom he seems to use as his touchstone of Christian believers here, should be ready to accept this nontranscendent reading of instantaneous conversion.

One strong argument for James's horizontal vision is that there is no clear mark that captures all converts, as all supernormal incidents can come about "by way of nature, or worse still, be counterfeited by Satan" (194). Indeed, in *Varieties*, as well as in "The Sentiment of Rationality," we see evil periodically lurking around the edges of James's thought. Leaving the satanic option outside his explanatory framework, however, he makes the case that all the marks of conversion, like "the real witness of the spirit to the second birth as found only in the disposition of the genuine child of God," can also be found in those who have no crisis and even outside of Christianity (194). James's thesis, therefore, which he finds substantiated in Jonathan Edwards—who seems to serve as a model American version of European Protestantism for James—but which in my view could also be substantiated in Emerson, is that "there is no chasm between the orders of human excellence, but that nature shows continuous differences and generation and regeneration are a matter of degree" (194–95).

If the difference between generation and regeneration is simply a matter of degree marked on a horizontal landscape of only one level, then we are faced with the question of what the storied universe stands for. Is there really "a redemption into another universe," as James previously stated, which indicates something beyond nature, or is there merely another perspective of nature and a differently conceived universe through faith? To put it in terms amenable to my larger trajectory: for James, does conversion require a different orientation of the self, and if so, is there a different conceptualization of nature that goes along with it? The question brings us back to Proudfoot's comments on James's moral universe and on religion as the relation to the unseen order. In the next section I will explore the connection between the self and the unseen order.

The Quest for Religious Naturalism 2: The Unseen Order and the Unseeing Self

To further elucidate the connection with the unseen order mentioned by Proudfoot, I want to take a different approach to James's *Varieties* by moving on from a focus on the converting or converted self, in relation to which we saw James in the last lecture dissolve traditional subjecthood into a field of energy, to a more capacious and active notion of selfhood that can reliably span this field. My approach is meant to help explain why James comes across as so generous in incorporating the various first-person accounts of conversion to the point of seeming unscientific and perhaps uncritical. To help explain this generosity, I want to invoke what the English novelist and poet Thomas Hardy terms the unseeing self. Hardy unfolds this idea in a revealing contemporaneous poem published in 1902, called "The Self-Unseeing." With the idea of the unseeing self, Hardy captures the dynamics by which the self is unable to experience something when seeing it but only has access to that experience when unseeing it.[38] After citing Hardy's poem, I will explain how the notion of the self-unseeing—which I take to be nonreligious but not unsuited for use in religious debate—can help us to close in on James's notion of selfhood, as well as on that of nature and the universe—that is, of the unseen order.

Hardy's remarkable poem was first published in his collection *Poems of the Past and the Present* (1902):

THE SELF-UNSEEING

Here is the ancient floor,
Footworn and hollowed and thin,
Here was the former door
Where the dead feet walked in.

She sat here in her chair,
Smiling into the fire;
He who played stood there,
Bowing it higher and higher.

Childlike, I danced in a dream;
Blessings emblazoned that day;
Everything glowed with a gleam;
Yet we were looking away![39]

In this powerfully concise poem, which stands on its own and is hence directly accessible, Hardy depicts an adult poet reflecting on what is presumably a childhood experience. He goes back to a familiar place, his childhood home, whose occupants—his parents—are absent. The precise geography of the home notwithstanding (here . . . there), which serves as another instance of spatialization, their feet are dead and the fluidity that is inherent in the passing of time, unlike the material solidity of the home itself, underscores that the past has indeed vanished. Through his memories the poet is able to bring his past and the house's former occupants to life, as he evokes an intimate family scene, with his mother smiling into the fire and his father, playing a violin, "bowing it higher and higher." Their mysterious and somewhat ecstatic aliveness results in the poet's trancelike dance, making him childlike—a sheen away from the real child that he once was—on a blessed day when all things shone brightly. Letting go of his skilled alliterations in the last stanza, Hardy ends the poem by retreating from the center of energy—the childhood scene, the day basking in glory—back to the margin. Reliving how even in that past, the people described were escaping, the poem wryly concludes: "Yet we were looking away!"

The poem's final sentence forms an *inclusio* with the title "the self-unseeing," thereby reinforcing its programmatic force of retrieving the memory of experience by vacillating between self-ignorance and unselfconscious immediacy. Together they convey a paradox that envelops the entire poem and holds its notion of selfhood up in indefinite suspension.[40] In the programmatic title of the self-unseeing we have at once a conscious acceptance of the past as closed off and an attempt to bring it back to life in memory, while the final sentence sounds a note of melancholy critique in that the child's unselfconscious joy and dance in the last stanza is precisely what makes it complicit in the demise of the past. As an adult the poet can only try to bring his experience alive by unseeing, but he cannot bring back the past itself, let alone revive those who inhabit it. As Peter Simpson plausibly states, it is as if the fire into which the mother smiles spreads from the center, spreading gleam but also doom insofar as it cancels out the past in the present.[41] On another, more Jamesian level, we might say that it is as if the center cannot hold; that is, the experience cannot last, and the dancing child is eventually forced to retreat to the margin, with the dream reifying into the hard crust of history, until the poet as an adult through unseeing becomes attuned to the experience again.[42] What the notion of the self-unseeing refers to, according to Mark Ford, is a sense that only through unseeing can the speaker work up the momentum to get at the experience of the past, which seeing only closes off but which unseeing opens up; but it does not do so indefinitely, for the final sentence, "yet we were looking away," forever seals the past into the poem. To formulate this in terms of a Jamesian conundrum, experience as a live option requires unseeing. Unseeing is that moment that the narrators of first-person accounts have captured, while seeing itself, the empirical and scientific thing to do, rather shuts out experience. The moment of "the self-unseeing" therefore allows Hardy to set up the poem as internally constitutive of experience rather than externally reflective of it. Far from a disappearing act, the moment of the self-unseeing is a de facto guarantee of self-presence, a commitment to a healing and lasting self-engagement.

If we return to James, I see the value of Hardy's self-unseeing in that it is what allows entry into experience as a field of energy without falling into the fallacy of either needing it to be scientifically accounted for, as James's references to the psychological research on automatism would perhaps suggest, or to be romantically drawn into, as the focus on conversion as self-surrender in the service of salvation may intimate. James avoids precisely this dilemma by letting the religious experience he is after be communicated through the literary texts that he has selected—often religious autobiographies or conversion scenes, not unlike Hardy's poem itself, in which the poet's childhood experience has been sealed and thereby made available. Rather than seeing the literary form of these *documents humains* as compromising the actual experience conveyed by them, as if they somehow make it inaccessible or cover it over, James is confident that, when looking at them as "thick" texts, they not only faithfully communicate religious experience but do so with sufficient clarity and gravity as to allow their readers and hearers—and one can imagine viewers, if one were to contemplate the role of film as a thoroughly modern medium—to get at the actual experience they describe. It is as if through the moment of unseeing, the experience itself has broken loose, while its power validates the literary shape in which it has become transmitted; the reason it can do so is because in the field of human consciousness, center and margins are inescapably connected.[43]

Along similar lines, the pragmatist James can also collapse the subconscious self as responsible for the metamorphosis of conversion into the orthodox Christian view of grace, whereby powers transcending the individual seize control of one's inner or underlying field of energy. It is the value of the forces at work, as determined by their effects, that motivates him. This question cannot be resolved by tracking down the origin or source of these forces, as origin is simply not what authenticates them for James.

Maybe it is that same relaxed attitude also, more than any personal or doctrinal commitment or ideology, that allows James not to let go entirely of what he calls "the admirable congruity of Protestant theology with the structure of the mind as shown in such experiences." The language he chooses in this context speaks of redemption as a free gift or nothing and of the gift of grace (199). The suspicion may linger that James clings to Protestant stereotypes by default, as, in his view, nothing in Catholic theology has spoken so directly to sick souls. Aware that dogmatism is better avoided, James summarizes his position: "As Protestants are not all sick souls, of course reliance on what Luther exults in calling the dung of one's merits, the filthy puddle of one's own righteousness, has come to the front again in their religion; but the adequacy of his view of Christianity to the deeper parts of our human mental structure is shown by its wildfire contagiousness when it was a new and quickening thing" (*Varieties*, 200).

While the blunt pragmatism of James's view here is abhorrent to modern sensibilities, especially insofar as it borders on historical opportunism, it accords with his oft-stated desire to judge a religion by its fruits rather than its beliefs, with the immediacy of the fruits observed here about Luther adding to their validity. Rather than merely adopting Luther's position as monolithically authoritative, however, James distinguishes between faith as intellectually conceived and "the more immediate and intuitive part, that is, the assurance that I, this individual I, am saved now and forever" (200). For James the state of "assurance" connects with the affective experience in which he is primarily interested. The state of assurance has a number of characteristics: first, the loss of all worry, and a willingness to be; second, a sense of perceiving heretofore unknown truths, which are sometimes even unutterable; and third, an objective change that the world undergoes, a sense of clean and beautiful newness within and without, in the language of Jonathan Edwards (201–2). Continuing his descriptions, James brings in again the automatisms to illustrate the newness of the convert's state. But with the exception of photisms (204), such as the light that Saint Paul and Constantine saw at their conversions, which he possibly highlights owing to their frequent occurrence, and the phenomenon that James describes as the ecstasy of happiness (206), he is actually quite clear that such phenomena can on the whole be ascribed to "the subject's having a large subliminal region, involving nervous instability" (204).

What James is after is not a backhanded way of bringing in the Holy Spirit, or secular traces of it, as he considers the above characteristics to be epiphenomena of the conversion itself. Whether these abrupt conversions are transient or permanent does not determine their efficacy or validity for him, since not even love is irrevocable. In a comment reminiscent of Emerson, he holds that a short conversion experience is also important, for it "shows a human being what the high-water mark of his spiritual capacity is" (209). Remember how in his "Divinity School Address," Emerson commented that dogmas and prayers "mark the height to which the waters once rose" (*CW* 1:87). This different application of what is essentially a very similar trope is noteworthy. For Emerson, the mark on the quay, as I called it in Chapter 1, indicates an atrophied tradition, one that produces preachers who do not give bread of life. The high point of Christianity for him is the sermon, the frank speech of man to men, displaying a decidedly horizontal thrust, if you will, and meant to energize and relieve us of the dead weight that characterizes dogmas and prayers. For James, however, conversions, as well as prayers and mystical visions, play an altogether different role. By showing a human being what the high-water mark of his or her own spiritual capacity is,[44] they are as much aspirational self-exhortations as they are traces of experience with great affective impact.

The self-reflective quality of James's comment on the high-water mark, as opposed to the impersonal voice with which Emerson speaks to us,[45] also reveals another effect of what I have tried to capture under the rubric of Hardy's unseeing self. That is, by connecting the external reflection or trace of experience (the high-water mark) with the internal integrity with which the converting person is driven outward into the world of action (the spiritual capacity), James forges a *direct* link between them in ways that prayers and dogmas can never achieve for Emerson. For Emerson, prayers and dogmas are by definition *Fremdkörper* (foreign things) in relation to the ordinary life and business of the people, while for James, conversions, however imperfect, short-lived, or literarily rendered, are as close to the fulcrum of religious experience as we will ever get. The fulcrum of religious experience here is the galvanization of humanity's field of energy, such that it affects the habitual center *in spe et in re*, so to speak, motivating one to move from aspiration to act. This movement causes a turning point that is, at the same time, a point of no return. Their transcendent cause or divine origin appears not to be the central issue, and ultimately their confessional background is not either. The only question that remains is that of the nature of the link between the psychological landscape and the unseen order.

What I have wanted to call out so far with Hardy's unseeing self is that the religious person for James is at once a trusting self and one that must perforce always look askance. Rather than choosing either to study or to undergo the transformation that conversion entails, setting divine off against human, Protestant against Catholic, or agent against object, James posits the religious person as a self squarely in the middle of a substantive transformation. Hence, he deems this self to be primarily *coping with* conversion, whatever it may be and however it may best be described. While looking for redemption and grace, the converting self is indeed seen as decisively— which I prefer, as a more pragmatic term, to "radically" or "empirically"— affected and deeply changed. Following my above conclusion, I think that the central issue as James sees it is not the alterity of grace, or the solace of redemption, nor can it be measured in terms of depth or shallowness, since James dismisses the need for such impersonal and nonrelational categories out of hand. Rather, under the guise of transformation, as if under a shield, it is the malleability of the sameness of life that conversion celebrates, the festive embrace of the old made new. In the Hardy poem the realization "yet we were looking away" enacts the self's unseeing by making the initial description of all that is evoked ring alive, hence true, even if its reality is closed off. So conversion for James also achieves completeness through the closure of the past that creates new life, the new life of a revived reality that has absorbed past memory, and thereby cleanses it. Maybe the most important phase of conversion stories is not the cresting moment of surrender—though James assigns that a high priority, and in narrative terms it marks a dramatic climax—but rather the slow buildup, the rising and falling of the emotions, the trembling of the psychological landscape before it becomes resettled. Precisely for that reason it is important for James to rely on a literary document, as it makes for a fuller, more material, and more lasting registration than any objective and impersonal readout of scientific data could give.

This brings me back to the final problem of the unseen order. Much has been made in recent years of the fact that toward the end of his career James believed in a pluralist universe and held "a pluralistically panpsychic radical empiricism."[46] Building on the above and leaving most technical aspects of his later philosophy aside, I want to draw some inferences related to it in light of my project of "thinking nature." James's enemies in developing the view of a pluralist universe, that is, the opposing viewpoints he needs to overcome, are absolutism and idealism, the latter represented by his Harvard colleague Josiah Royce. One way in which James goes about doing this is by relinquishing the bifurcation of established philosophical terminology into *rational* and *irrational*, replacing it with the terms *intimacy* and *foreignness*. His aim, which was already clear in "The Sentiment of Rationality," is to be more intimate in and with the universe, for which he is willing to pay the price of upholding a view that is not absolute but is able to accept and accommodate evil and error. This intimacy certainly affects his view of the divine, which he sees as finite rather than infinite. It makes him ultimately perhaps less the desired Christian thinker on modern religion than Charles Taylor might have preferred but a better role model in the end for Joas's position.

Ultimately, James moves to an embrace of a superhuman consciousness, which David Lamberth has plausibly interpreted in connection with the notion of the wider self already mentioned in *Varieties*, with which James seems to want to reach beyond the mere sum of the conscious and the sub-conscious self. Be that as it may, yet careful to avoid any absolutizing aspects of the superhuman consciousness that he envisages, I wonder whether what we see in James's pluralist universe is not also a coping mechanism at work—that is, a case of the self-unseeing on a cosmic scale, in which we are able to contemplate the absolute in its environment, as James calls it, as a center with ever-expanding margins. In adopting the notion of a cosmic coconsciousness through a bond marked more by intimacy than foreignness, James can keep alive the duality of a subject that is trusting and looking askance at the same time. Whether this amounts to pantheism is a question that may be hard to answer, as James clearly sees bonds of connection everywhere in the universe and calls these out as social relations in the service of the greater cosmic intimacy he desires. What may be more fitting to say about his universe is that, by moving as far away as is humanly possible from solipsism, James has evoked a cosmos in which human beings are meant to feel not just at home but truly welcomed.

(*Thinking Nature*) . . . and the Nature of Thinking

I know that the world I converse with in the city and in the farms, is not the world I *think*. I observe that difference and shall observe it. One day I shall know the value and law of this discrepance. But I have not found that much was gained by manipular attempts to realize the world of thought.

Ralph Waldo Emerson, "Experience"

This book has been about the world of thought and nature, thought of nature, that is, or, more precisely and dynamically, it has been about "thinking nature." As Emerson says in the motto quoted above, the point has not been to realize the world of thought or even to judge thought by whether or not it could be realized. There is an integrity to thought itself, and in the case of "thinking nature," there is an integrity to the concept of nature as it is thought. These three different but interrelated themes, the integrity of thought and the integrity of nature lying behind it, which in turn inspire and drive "thinking nature," are what I have wanted to bring out.

In Christian thought, which is how we can classify many if not all of the sources that I have analyzed, we ask too quickly the question of whether something is orthodox or heterodox, whether one premise connects neatly with another, whether creeds are checked off and doctrines heeded. This hypervigilance turns thinking into a tight mathematical puzzle that can stifle the imagination rather than unleash it, silence poetry rather than set it free or ablaze; it is monologue rather than conversation, and, most of all, it fails to let the object of thought drive the process whereby it is thought. It is my contention that owing to our default instinct to impose prefabricated and artificial hegemonic categories of thought on the object of thought, religious thought in particular and, more narrowly, Christian theology has suffered. In particular I want to point out that Western religious thought has been remarkably consistent in retaining what I would call a scholastic character, inoculating itself against criticism by putting itself beyond experimentation and also beyond experience.

When scholasticism developed in twelfth-century Europe, it was a school method that was functional and meant to guide the classroom practice it reflected.[1] What has been far less realized is that it resulted in a makeover of the dynamic, guileless, and literary patterns that had shaped the Christian thought of an earlier era. That makeover was so drastic that to this day we have great difficulty retrieving the earlier outlook.[2] With all the radical intellectual developments and religious incisions that have brought us from the Middle Ages, through the Reformation and the Enlightenment, to postmodernity, it is amazing that in theology a certain scholastic attitude has not only survived but prevailed. Be that as it may, the follow-up question is whether the survival of this theological mode has also *invigorated* theological thought. This I venture to doubt, and I have tried to make the case that there have been successes but also failures as far as "thinking nature" is concerned. What this book has argued is that cases of "thinking nature"— such as Eriugena and Emerson, who are the antitheses of scholastic thinkers—should not be judged on the basis of orthodoxy, nor of confessional, ecclesial, or even academic fit, but by the integrity with which they carried out their own, or nature's own, project. I have also argued that, if indeed we find a subterranean tradition of "thinking nature" running through Western religious thought, that tradition is, first of all, not a countertradition, for it is rather opaque and intertwined with other traditions. Second, because it is not easy to isolate, it should be taken very seriously as a resource rather than targeted or suppressed as a distraction.

Given the perilous centrifugal state of what is considered mainstream Western religious and theological thought in the academy today, a state that is only heightened where nature is concerned, we simply cannot know when such a resource may come in handy. I would like my mapping out of this tradition to be there for the taking for whoever wants it, for whoever wishes to bring a more imaginative approach to religious and theological thought to the fore, regardless of if or when the hold of scholasticism eases. Put differently, I offer these readings to the imagination, paving the way for a time when a new model for religious and theological study can lift the byzantine constrictions of their respective orthodoxies.

The Dynamics of Premodern Nature

At the end of my project on "thinking nature," then, I want to evaluate the impact of the foregoing chapters and discuss where the exercise of "thinking nature" has left us and what I see as the potential consequences for the "nature of thinking"—that second part of the title, which will here take center stage. Given my focus on religious thinkers, it is no surprise that my comments will especially touch on the consequences for religious thought. While I will end by grouping my findings around the figure of Emerson, whose galvanizing presence first inspired me to develop the cross-historical axis of "thinking nature," I will begin by commenting on the book's two-fold structure.

My approach at the start of this project is clear enough: the robustness of Eriugena's thought makes him an outlier in the sphere of medieval studies, as well as in the history of philosophy and theology, to the extent that he is considered either a heretical footnote to an orthodox past or, at best, a pantheist exception to what is otherwise construed as a cohesive premodern tradition in which nature seamlessly overlaps with biblical creation. As I propose, reading Eriugena through Emerson, the latter being an author similarly engaged in "thinking nature," better prepares us for what I see as lying at the heart of his thought: the encounter with nature in its otherness, that is, as a diffuse, imaginative, grace-suffused whole, crisply separable neither from the divine nor from humanity. It also brings out that, for Eriugena, nature is dynamic and relational, while at the same time inherently geared toward the divine.

Given that the bond of causality seems too thin to channel Eriugena's unwieldy premodern nature, I place considerable emphasis on nature's mission of bringing humanity back to God, as return has always played a key role in the connection between humanity and God. In premodern Christianity, more than the mirror image of procession (the movement by which creation flows out from God), return carries the overtone of restoration and even conversion, imposing a moral and religious imperative. This places a heavy burden on human nature because it follows that if human beings are the image of God, then they need to work toward God from within nature, in which they are integrally embedded. Supplementing causality, the kinship of the *imago Dei* and the divine gives humanity a richer, more comprehensive bond with the divine, the strength of which can absorb and even override ordinary causal ties.[3] When the causal bond recedes, a set of thicker, richer, and what William James will later call more intimate forms of connection come to the fore relating humanity to God but also humanity to creation and creation to God. Given its inherent expansiveness, it falls to nature to navigate these various connections and to integrate them within a single, nondivisive whole.

It should be no surprise that, in premodern Christianity, the figure of Jesus Christ is a catalyst for working out the respective relations encompassed by the universe. Maximus the Confessor's remarkable cosmic Christology, spread by Eriugena to the West, has long been known for doing just that.[4] It includes a number of insightful and deeply original ideas, notably that of the human person as the workshop of all things (creatures), of incarnation as incrassation, a kind of material fattening, and of nature and scripture as the two garments of Christ at his transfiguration, a proleptic moment of eschatological cosmic glorification. Making Christ the lynchpin of his cosmic liturgy, as Balthasar called Maximus's theological project, the latter transforms universal return into a more intimate narrative in which Christ's role is as dynamic as it is central. By redeeming humanity, the workshop of all things, Maximus demonstrates that Christ redeems the entire cosmos. Conscious of Origen's perceived depreciation of material reality, Maximus casts redemption not as compensation for the fall but as adding luster to a universe that, in the tidal wave of Christological beneficence, is destined to become the divine's eschatological home.

For Maximus, Christ does not only sanctify nature at redemption but already at the incarnation. For by assuming human nature, Christ assumes de facto all of nature, which is underscored by the material notion of incarnation as incrassation. Yet by anchoring redemption, as an extension of Christ's incarnation, in humanity's created, natural state, Maximus sows a seed that causes problems later on. Insofar as redemption equals return, it may seem that the ungendered, eschatological state Maximus ascribes to the resurrected Christ, based on Galatians 3:28 ("in Christ there is neither male nor female"), is an effective way to heal the gender division of created, and fallen, Adam and Eve. But while it is in the power of the risen Christ to effectively transcend humanity's gender division, Maximus's desire to sustain the redemptive Christological effect of his cosmic liturgy makes him subsequently latch on to the church's monastic office, as salvation by the risen Christ is henceforth continued in the prayers of male ascetics in the church.

My criticism of Maximus's cosmology here may seem unfair, for how can we hold his ecclesial circumstances responsible for the limitations of his theological profile? I submit that, indeed, we cannot. Yet the deep admiration I have for Maximus does not make his view that the risen Christ overcomes the gendered creation of humanity less of a theological roadblock. Let me explain. Transcending humanity's gendered creation in order to neutralize division implies that Christ's redemption—at least in its first stage, the reconciliation of the gender division—entails the work of denying that division. But the effect of that same division is paradoxically exacerbated when the continued effect of cosmological redemption is made dependent on the prayers of male ascetics, insofar as they are tasked with continuing Christ's redemptive work. To the extent that earlier thinkers like Gregory of Nyssa and Ambrose also see the resurrected Christ as having neutralized and overcome creation's gender division, their Christological solution is likewise both ingenious and theologically flawed.[5] As long as Christological solutions take their ultimate cue from the resurrected Christ as having overcome gender division and, from him, not only circle back to the male Adam as marking the ideal human state before the Fall but also, in a next step, project it onto male ascetics, they can never be fully effective—that is, equally redemptive of all of humanity.

If this were merely an issue of ecclesial circumstance, it could be dismissed as an atavism awaiting an institutional update. By tying Christ's redemption through incarnation directly to humanity's created, divided state, seeing the fall reflected in the gender division that is subsequently overcome in the resurrected Christ, Maximus fails to make redemption fully effective for the cosmos in that it necessarily depends on that gender division. Aware that history affords us an interlude before the actualization of cosmic liturgy into eschatology, Maximus is cautious not to take ecclesial liturgy as a direct substitute for redeemed creation. This allows us, instead, to widen the circle of ecclesial liturgy to match that of *natural reality*.[6] As I argue in my chapter on Maximus, the imperative of return would seem to demand that the two circles eventually overlap in full.

There is a further point to be made. By criticizing Maximus's concept of nature, I not only take aim at his premodern theology but also at theologians and historians of theology who have taken up the study of premodern patristic giants like him. The current popularity of "la nouvelle théologie," a French mid-twentieth-century Catholic movement of intellectual renewal that put the spotlight on a select group of Platonic patristic thinkers, helps to illustrate the problem I have identified. Admittedly, the historians of "la nouvelle théologie," especially Henri de Lubac, Jean Daniélou, and Marie-Dominique Chenu, had good taste; they selected worthwhile thinkers, even if their choices exhibit a degree of historical datedness. As studied by Balthasar, who was himself on the margin of this movement, Maximus clearly belongs in this company of patristic heavyweights. Yet we should not overlook the fact that the aim of the "new theologians" was to mobilize the patristic and medieval past for an internal renewal of the church in the present, not to take off any critical edge by freezing these thinkers in time. Yet, since various patristic authors have become largely known to us through the explorations of the "new theologians," there is the risk that we fail to do our own critical explorations of these sources, which is what my project of "thinking nature" aims to do. Since the mirage of a respectable orthodox, but immutable, past seems to draw many contemporary theologians to patristic thinkers today, I use the exercise of "thinking nature" to point to a flaw in Maximus's theological thought that has been overlooked by Balthasar and others but should be taken into account today, even as my criticism is meant to be a faithful appraisal. That flaw in Maximus concerns the details of his Christology as the principle of cosmic redemption.

The evaluation of Maximus allows me to propose a first benefit of Eriugena's "thinking nature" over Maximus's better-known Christological cosmology. Far from an iconoclastic move, Eriugena's project of "thinking nature" does not detract from Christ's cosmic efficacy, a position that Eriugena adopts from Maximus. But by encapsulating it in a larger whole, the *Periphyseon*'s "thinking nature" reflects more faithfully the power of Eriugena's own medieval thought as a deeply inclusive affair. "Thinking nature," in the sense in which he takes *natura* to be comprehensive of all of reality, liberates us quite literally from the binding strategy whereby gender division is a scourge to be overcome in order to make the puzzle that is Christology work.

Treating Maximus before discussing Augustine allows me to bring out that the ascetic imitation of Christ, tied as it is to the reunification and redemption of the cosmos, reflects by and large an Eastern rather than a Western pattern. Inasmuch as Western Christianity was marked by asceticism, its quality was initially Ambrosian and Mariological, even though Ambrose's thought gained paradisiacal overtones later on. After Augustine, Gregory the Great made asceticism a matter of exegetical discipline. That is, Gregory tied the struggles of ascetic life closely to the moral life as filtered through tropological exegesis. He thus came to see the ascetic life, with reference to an expression from novelist Robert Musil, as a way "to live as one reads"[7] rather than as primarily the result of social or personal choice. Gregory's changing modes of reading shaped a new medieval culture of embodiment. In light of his reigning tropological model of medieval embodiment, Augustine seems to fit the mold of "thinking nature" surprisingly well. Even as he adopts an ascetic posture for himself, he by no means implies that asceticism should be the normative lifestyle for all Christians. Hence, he cannot be so easily seen as the repressive thinker he has been made out to be. If in terms of church leadership he betrays the same clerical constraints as Maximus, these do not affect his take on "thinking nature." Based as it is on his literal reading of Genesis and concomitant embrace of creation in its independent reality, his views are not primarily Christological. Given his defense of marriage, one can almost imagine him not being an ascetic. Even the opposition of grace and nature does little to make him critical of creation in ways that one might perhaps have expected.

This is not to say that Augustine's thought resembles Eriugena's. Far from it. Yet what they have in common is the joint dependence on scripture as not just an anchor but a moral guidepost that propels them forward on their intellectual journey. No modern thinker that we dealt with in the book's second half—be it Schleiermacher, James, or Emerson—betrays a similar commitment to scriptural truth, which seems to have simply disappeared in post-Reformation modernity. While at least for Schleiermacher and, to a lesser extent, Emerson, scripture remains a source from which to draw, it is not a yoke to shoulder, a treasure to hunt, making theologians into exegetes always ready to cast out their fishing rods. The prominence of scripture may well seem a commonplace, for what early or medieval Christian thinker would not rely on scripture? But there are few medieval intellectuals who breathe scriptural air as unassumingly and yet as commandingly as Augustine and Eriugena do when it comes to nature. While in many a contemporary Christian environmental account the theme of stewardship holds pride of place, the fact that, for both these thinkers, nature or creation can be encountered wholesale through scriptural engagement bound them together long before the stewardship model took root. In Augustine and Eriugena, the process of reading scripture, especially Genesis, sets nature on a path that highlights its abiding newness and irreversible material reality alongside its endemic promise of restoration.

With the antignostic sting taken out of *creatio ex nihilo* by Augustine's time, creation in both authors is further marked by temporality. In Augustine nothingness is something to be staved off, as nature moves away from the abyss, with the days of creation marking an irreversible series of forward steps whose importance is reinforced by his choice for a slow, literal interpretation. In the more allegorically inclined Eriugena, in contrast, nothingness betrays the apophatic eminence of the divine itself, whose superessentiality harbors any and all possibilities, which, in turn, stand ready to take on a life of their own once they are teased out and actualized. In both Augustine and Eriugena nature symbolizes a life of one's own in the face of the divine, with creatureliness not a watered-down version of divine life but a divine exhortation to explore new avenues, to chart new and untrodden territories.

Contrary to their perceived reputations, according to which Augustine is the champion of orthodox creation and Eriugena the pantheist outlier, all this seems to align the approaches of Augustine and Eriugena. That their divergence is less striking than previously thought is evidenced, for example, by their joint preference for a hexaemeral account, even if Eriugena's thrust is more allegorical. *Thrust* is indeed the operative exegetical term here, and it is the sense of an active forward motion that motivates their deliberative exegetical choices, lending them a degree of Emersonian onwardness.

Which way do we go? Which path do we follow? Instead of the meandering ruminations of Gregory the Great, or the panoply of erudite choices espoused by Jerome, the path is never so overgrown with exegetical foliage in Augustine and Eriugena that a course cannot be detected; the tunnel is never so narrow that there is no light beckoning them forward. The more they are beckoned forward, the more each of them allows not only for humanity but for all creatures to raise their own voice. This is an important point in the book's first part, the idea that creatures do not only have meaning as signs, with natural signs read and interpreted through the verbal signs of scripture, but that they can raise their own eloquent voices alongside and in concert with the words of scripture. Nature is thus a soundscape of earthly polyphony. With creatures raising their own voice, humanity is no longer in control of the conversation or even its sole interpreter. Rather, humanity contributes to nature's onwardness much as creatures do—that is, as a valued interlocutor, as selves that are part of a bigger whole.

To see nature as the eloquence of things according to Augustine, from whom I derive the idea, and as conversation in Eriugena, into which I develop it to allow for a dialogical back-and-forth on multiple levels, helps us to make sense of all the lulls and gaps, the wide-ranging digressions and excursions that we see in Eriugena's *Periphyseon*. Rife as the dialogue may be with repetitions and digressions, a sonorous voice nevertheless holds its different parts together; this sonorous voice grows in strength as the digressions are strung along as beads in a necklace. I have increasingly come to see each discrete unit of the *Periphyseon* as necessary in the conversation that is nature, whose purpose it is to achieve greater adornment for its collective enterprise—namely, to give the cosmos, through persuasion, a more powerful presence by leading it from beginning to end. Of course, nature in Eriugena is qualitatively different from creation in Augustine, for Eriugena's nature emphatically includes God. Surprisingly, however, this does not appear to set Eriugena's discussion that far apart from Augustine's. The difference is more of degree than of kind. If we can agree to see things this way, then the pantheism charge dissolves, for God also becomes another interlocutor in the earthly polyphonic soundscape that the *Periphyseon* presents.

So far, so good perhaps, but I have obviously presented Eriugena's project as one on "thinking nature," not on nature as conversation. What, then, do I see as the advantage of "thinking nature" over the model of nature as conversation? I have highlighted conversation in order to bring out the encounter with nature, of humanity with nature but also of various aspects of nature with each other, preferring encounter to instrumentalization or other forms of use. Encounter, in turn, acknowledges nature's agency. I have also highlighted conversation in order to allow for nature to be presented as a consonant but nonmonolithic whole that can meaningfully account for and factor in gaps, odd silences, and digressions. What the process of "thinking nature" does is to insert the element of self-reflection as constitutive of nature's identity. Nature's self-reflection is what allows Eriugena to complement and correct Maximus on the point of his Christology; it also allows him to bring more focus to the early Christian and medieval exegetical process, integrating its allegorical musings on a deeper level.

Eriugena's project of "thinking nature" leads him to include God in nature, not as a way of domesticating the divine but of furthering the proximity of collective nature to God, turning proximity into integration. The result is not a denial or downplaying of humanity's created nature in favor of a generically divine universe; the emphasis on self-reflection would appear to prevent this kind of pantheism-fallacy. Rather, because self-reflection, through its inherent relational quality for human beings created in the image of God, extends naturally to the human self's increased engagement with God, Eriugena can cast "thinking nature" as a fully religious project to which humanity and the divine are jointly and equally committed.

A few comments on Eriugena's distinctiveness from Maximus and Augustine will round off this section on his premodern project of "thinking nature." Eriugena initially adopts Maximus's views of Christology; he elaborates them further by bringing in Gregory of Nyssa. Yet, however Maximian Eriugena is, the innate freedom of "thinking nature" precludes his committing himself to it and to the gender denial implied by Maximus's risen Christ and male monks. Eriugena's somewhat muddled solution of giving a superadded body (*corpus superadditum*) to the gendered Adam and Eve, one that will be stripped off at the end of time, is an awkward way of getting around precisely the kind of problems found in Maximus and Nyssa, but it should not be judged a depreciation of embodiment. His definition of humanity as an eternal notion in the divine mind has been assessed in scholarship as evidence of idealism, even as it is offset by the Maximian emphasis on Christ's incarnation as incrassation, an embrace of material reality to the fullest, which Eriugena carries forward into his own work. Where Eriugena fundamentally diverges from Maximus, it seems, for I want to tie the above examples together, is in his willingness to radically follow the course of *natura* in all its meanderings and limitations. Acknowledging that human beings are indeed flawed, or weighed down by sin, and that their rational judgments do not live up to the ethereal height of their status as primordial cause in the divine mind, Eriugena can then turn that human deficit into something that accords humans their firmly rooted, natural, material place. In sum, the focus on "thinking nature" helps to show that anthropology is not an imperfect substitute for Christology, that giving it pride of place does not detract from the Christian nature of his thought.

Rather, it shows that human beings are sufficiently well-equipped through dialogue, not just among themselves but with God and with nature, to shepherd nature from its inception in God to its return.

In my view the *Periphyseon* is the natural theology that Augustine never wrote. That characterization underscores how much I see Eriugena's medieval thought as being firmly anchored in the Western tradition. Following Augustine's commitment to scripture as a guide, it appears that Eriugena builds a theology of nature that, while being deeply exegetical, is focused on constructing a discursive intellectual path of procession and return, long before and more exegetically grounded than that of Aquinas. In this way Eriugena's *Periphyseon* simultaneously affirms and deeply troubles the judgment of de Lubac that theology before scholasticism is identical to scriptural interpretation.[8]

The Dynamics of Modern Nature

It is clear, though perhaps regrettable, that scripture is no longer a major factor for the modern religious thinkers whom I have engaged in this project. Few post-Reformation theologians have been consistently and structurally committed to charting their work along an exegetical course.[9] This change can be detected even prior to the Reformation, for the case can be made that even for scholastic thinkers such as Aquinas, scripture is a source to be mined rather than a path to be followed. This would make the Reformation itself a kind of exception, as it was soon followed by Reformed scholasticism.

Be this as it may, there is a further reason why the modern thinkers in this book keep their distance from scripture. Rooted in the Enlightenment, they all face the problem of how to position themselves vis-à-vis a Christian tradition that is no longer perceived as coextensive with what is considered the trajectory of the intellectual tradition more broadly. The discussion between Hans Joas and Charles Taylor staged in this book's chapter on William James is reflective of the shift from a Western intellectual tradition seen as religiously inflected to a more secularly conceived one. For Joas it is important that we be able to produce scholarship in the history of religions in a way that allows for our personal beliefs or commitments to be part of the equation. This would be a de facto break with the tradition of this scholarly field, as well as with the Enlightenment mind-set more generally, in which it is rooted. But this break is innovative and may even be desirable. To have a paradigm whereby the history of religions disqualifies personal commitment, and theology presupposes it, requires premising the study of religion on what I consider an irrational—or foreign, as James would say—polarity. For Taylor, William James is an attractive point of departure for the study of religion precisely because, oblivious to the above polarity, he appears to speak to those who find themselves on the cusp—that is, neither inside the religious fold nor by definition completely outside it. Whether a Jamesian approach to the study of religion allows for the kind of faith commitment for which Taylor, admittedly not a theologian, seems to be seeking is a problem that we will leave unsolved for now. Still, it is Schleiermacher, not William James, who is considered the father of modern theology, and it is with him that I start my foray into modern examples of "thinking nature."

In view of his reputation as the father of modern theology, Schleiermacher's early *Reden über die Religion* is remarkably cosmically oriented. There is a sense that God can be encountered in nature, though it is less clear if nature itself can or should be encountered aside from or outside of God. Schleiermacher, under pressure from the constraints put on religious thought by Enlightenment philosophy, is too involved in salvaging the search for the divine to indulge in valuing such an extradivine encounter. In a time when the cultural paradigm that is religion itself is placed under strict rational scrutiny, the risk of losing this battle is too high. Hence, we have the odd situation that, on the one hand, we see Schleiermacher reach out to the cultured despisers of religion, the audience that we would increasingly come to associate with those interested in history of religions, while, on the other hand, we see him in *The Christian Faith* retreat and turn again to the tradition of old. For Schleiermacher this is a Reformed theology that is more tightly organized than before but not fundamentally reoriented, though it is no longer primarily scriptural.

There are two related points that I want to take away from the discussion of "thinking nature" in Schleiermacher that seem to apply to him more broadly. First, it is hard to gain access to Schleiermacher himself without somehow coming at him from a Barthian perspective or without carrying at least a sliver of the burden of Barth's neo-orthodox criticism, even if one is not won over by it. The two thinkers have simply become intertwined, which does a disservice to Schleiermacher. Second, and related, this entanglement has also influenced scholarship on Schleiermacher. Schleiermacher may have been rediscovered of late—and the more Barthianism fades away, the more he can be read on his own—but his reputation as a liberal still haunts him in that he seems forced to make the case of orthodoxy on his own, always battling the ghost of liberalism. In this respect it seems the way in which Schleiermacher scholarship has pressed on causality as the great bond between creation and the divine is not so much artful as artificial; one cannot quite suppress the sense that the *Speeches* breathe a richer air of nature, not so much liberal as lying in wait to be liberated.

As I point out in Part 2 of this study, I consider the connection between Schleiermacher, stripped of the Barthian lens through which he is interpreted, and William James as a more fruitful one than between Schleiermacher and Emerson. Whereas the latter connection forces us to compress and read Emerson too much in Schleiermacher-reception mode, the former can help us bring out better the focus on religious experience as a modern innovation, one that emerges and is embraced by both thinkers, albeit differently.

As Martin Jay points out in his *Songs of Experience*, experience came to the fore in modernity as a way to complement knowledge in a more secure way, which was needed given the increasingly asymptotic and regulative value of Enlightenment knowledge.[10] Experience would come to play an important role in the nineteenth century, the era for which Schleiermacher's *Speeches* of 1799 set the tone in many ways, especially in religion and aesthetics. One of the prime ways experience is deployed by Schleiermacher is in his struggle with Kant's practical reason, which puts morality over religion as constituting the more universal law. Schleiermacher is clearly interested in finding a way to write an apology for religion that does not telescope it into morality. Even though Romanticism is probably whence the cultured despisers hail whom Schleiermacher addresses in the *Speeches*,[11] Jay colors in the meaning of experience further for Schleiermacher by pointing to the Pietism in his background. While Schleiermacher marshals Pietism and Enlightenment thought to help rebuke the academic scholasticism that marked post-Reformation philosophical and theological thought, Jay sees Schleiermacher and James tapping into the reservoir of experience as a way to carve out a space for a religion reduced neither to some other intellectual or sensible realm nor to morality. Furthermore, he sees both aligned in the great emphasis they place on emotional intensity, though their respective European and North American contexts make them decidedly different in how they do this. Their thought exhibits idiosyncrasies that make it both irreducibly modern and irreducibly their own.

The above leads me to want to tease out Schleiermacher's Christology here, as I do with Maximus. For Schleiermacher Christology is no longer part and parcel of return as the flip side of procession but seems instead to lay bare a more narrowly anthropologically centered, deeply soteriologically motivated theology. As he aims for a kind of universal restoration, Schleiermacher does so without notable structural ties to scripture or creation, which is, as we have seen, the favored combination of the premodern thinkers engaged in "thinking nature." At times, it seems as if Schleiermacher moves on a dual track. The *Speeches* breathe a kind of cosmological climate, which is motivated but also enhanced by his desire to anchor the infinite in the finite. That would seem to make nature or cosmos an excellent and expedient double liaison, as nature represents the earthly realm but through its more Romantic overtones can also be seen to wrap itself in a more mystical cloak. Perhaps it is the endemic opacity of such an ambiguous role for nature that makes Schleiermacher decide not to explore that route in the end.[12] The institutional factor should also not be underestimated, by which I mean the combined weight of the church and the academy. Ultimately, Schleiermacher seizes on the incarnation as Christ's rootedness in and embrace of human nature and, even more, of history. This becomes the primary way that he interprets Christianity's excellence and teleological superiority over other religions. But while incarnation links Christology to nature through human nature, salvation must be navigated within the opposition between nature and grace, which in Schleiermacher, who is continuing the Reformed legacy here, must take priority. Of course, this leaves the problem of nature, which with Christ's role taken out is reduced to excarnate objecthood, linked to the divine through thin, unfleshly bonds of causality.

By sacrificing the role of nature as double liaison, Schleiermacher not only detracts from its meaning for modern experience, in the end perpetuating Pietism more than Romanticism, but also reins in the hermeneutical resonance of nature. "Thinking nature" could have been a way for him to continue a role for nature that would transcend delimited objecthood, even as he avoided the specter of Spinozism and all that it stood for. After all, Schleiermacher finds himself at an unusual but important historical crossroads, observing a renaissance of Greek ancient thought. Werner Jaeger has called him "the Winckelmann of Greek philosophy," thereby pointing to his enormous importance for the modern Plato-reception.[13] But with the great influence of Winckelmann on the reception of Greek art and thought in eighteenth- and nineteenth-century Germany, tilting it in the direction of neoclassicism, there emerges the awareness that Christian culture is deeper and richer than the Reformed Christian tradition with its patristic and biblical roots. Christian culture, rather, marks an inherent departure from an older civilized world that, in the final analysis, remains untouched by the Christian culture that broke off from it. Of course, the classical past must have a relation to the Christian past, but what kind of a relation precisely is unclear, just as it remains unclear how we should link Schleiermacher as "the father of modern theology" to Schleiermacher as "the Winckelmann of Greek philosophy," with both audiences being too far apart even then, and much more so now, to truly care about a rapprochement. Still, one wonders to what extent Schleiermacher's religious posture should be deemed apologetic, moved and motivated as he was by a deep awareness that Christianity was itself under siege and that to maintain its footing it should not just reorganize but retrench. Shrinking theology to the circle of anthropological salvation as its main theme may have initially seemed a triumph over other religions; however, it is in other ways an act of defeat vis-à-vis the inroads made by neoclassicism. On the heels of the latter movement religious scholars would soon begin to take a more equitable look at the entire spectrum of religions, while Christian thinkers would feel increasingly compelled to distill and analyze the essence of Christianity. In hindsight Schleiermacher was right to want to address the cultured despisers of religion, and it might have been a more commanding view, in light of the foreseeable future, if he had kept up his "either/or" conversation of the *Speeches* with that audience, prodding them not to abandon religion but reformulating its meaning beyond formulaic church doctrine.

At this point I want to come back to Schleiermacher's view of Christianity's doctrinal history, especially the fact that he regards creeds as the collective experience of the faithful.[14] On the one hand, this view adds a keen psychological perspective to which one can imagine someone like William James relating. On the other hand, it proves to be a weak historical position in that it does not seem to allow for temporality to leave its mark. Insofar as the importance of historical interventions, such as conciliar creeds, is overridden by the overall linear force of the Christian tradition, the situation of doctrine is not all that different from that of nature. In Schleiermacher nature also appears to have been frozen in time, the background canvas to a drama about salvation from which anthropology and Christology are increasingly isolated as driving his concept of religion. In both cases, that of doctrine and nature, the problem lies with Schleiermacher's view of the uniqueness of Christianity as expressed in a reconfigured relationship to the Old Testament and hence to creation. As is widely known, Schleiermacher relegates the Old Testament to extracanonical status, a position in which he is later followed by Harnack. While he wants to take the results of biblical criticism seriously by seeing Christianity as a unique religion that is not simply a continuation of Judaism, his view makes for a radical break not only with premodern Christianity but also with the Reformation, for which such a position would not just be untenable but unthinkable.

The premodern creation narratives we studied in this book do not appear to be interpreted via the idea that the Old Testament foretells Christ. This is not to say that, implicitly, this is not hardwired into the paradigm of our authors, but it does not seem to drive their exegetical decisions. Rather, the theme of creation is simply too substantive to be ignored, and in order to analyze it, the book of Genesis is the go-to text. The paradigm of premodern authors leads them to see the Bible as one whole, not necessarily divided into two parts—that is, as one book containing a plurality of voices and perspectives. Focus is then honed in a variety of ways along a variety of lines.

With Schleiermacher's foregrounding of Christianity's interruptive uniqueness, a different understanding of Christ comes into view, unrelated to the Old Testament. Thus, he considers Judaism "a dead religion," deeming the "original intuition of Christianity more glorious, more sublime, more worthy of adult humanity."[15] Not only is there no longer the possibility of a cosmic Christology, but, with the emphasis on the radical newness of Christ, as opposed to his role of a second Adam or a second Moses, the most credible connection with his followers becomes their faith in him. For Schleiermacher this faith takes place in a direct bond of their consciousness with Christ's full, sinless God-consciousness. Pantheism and thin causality may be the ostensible problems of Schleiermacher's way of "thinking nature," but underlying them is not so much an unhistorical Christ as one who is made to hover above history even if somehow springing from it.

This is not to say that Schleiermacher does not also see Christianity as "more deeply penetrating into the spirit of systematic religion and extending further over the whole universe."[16] But it is clear from the above that "thinking nature" is an increasingly difficult task for Schleiermacher in light of the extracanonical status of the Old Testament and the subsequent demotion of creation as a prime (biblical) theme.[17] The contrast with premodern thinkers is stark, then, given that they derive support for their own effort to "think nature" precisely from the Old Testament. If we look at this tangled nineteenth-century history, it is even more surprising that Karl Barth does not see a need to reestablish natural theology by reevaluating creation from a biblical or patristic angle. The enduring focus on redemption as a fideistic soteriological, meaning anthropological and Christological, locus seems precisely to have prevented this.

The fact that Schleiermacher stresses *Leben* (life)—seeing it as the source for experience described as *Erlebnis* (experience, related to *Leben*) rather than *Erfahrung* (also experience, which later becomes the dominant philosophical term)—contributes to the fact that his emphasis on experience is not in sync with the way the concept develops later in Germany.[18] But it is unmistakably clear from William James, and evident most of all in his influential *Varieties of Religious Experience*, that experience is there to stay as a central concept in modern religion. James's views are clearly not beholden to biblical criticism, which in the United States does not seem to be as closely intertwined with the course of theological study as it is in Germany. After writing his radical and impactful *Speeches*, Schleiermacher pursues academic theological work, including developing a theological curriculum, rather than designing a model for the study of religion more widely. This intellectual trajectory follows from his view of the supremacy of Christ's grace-filled consciousness. In contrast, James is much more interested in the latter, as commentators such as Proudfoot and Taves have both made abundantly clear, though the predominance of Christianity remains. Experience, for James, provides the warrant that the individual in solitude does indeed have a relation to God. But it also betokens, at a higher level of emotional intensity, the kind of self-surrender entailed in salvation on which James places considerable emphasis, as this experiential focus is what makes him draw out and dwell on the theme of conversion. In the new dynamic he sets up, Christology plays no role but creation and nature all the more.

Yet before we conclude all too rashly that James's only concern is the academic science of religions or is seen entirely as a forerunner of the contemporary philosopher of religion, it is relevant to point to the role Jonathan Edwards plays in the background of James's thought, as James had a particular fondness for him. Proudfoot has compared them insightfully, and I want to highlight some of his findings here, though I will put them to slightly different use.[19] I want to bring in Edwards, in part, because it has proved all too easy to remove James and Emerson both from the larger scene of North American theology, deeming them freethinkers who had no visible ties to the American ecclesial landscape. James's absence in the various American theological genealogies[20] makes his unfazed treasuring of Edwards only more remarkable. It is another sign that he pursues the agenda of his own concerns, even when those concerns are deemed the province of fields "not his," such as theology. It is one of the remarkable traits of James's keen mind that he does not care about such petty differences. In this he defies the attitude that so often characterizes the contemporary scholar of religion who would want to keep any association with ecclesial practice or institutional loyalty at arm's length.

Proudfoot argues that whereas one might suspect Edwards of highly valuing the place of introspection and see James the scientist as being more guarded and suspicious of it, in actual fact there is a skeptical side to Edwards, who is known to have been influenced by John Locke in developing his concept of religious affections. Wary of translating Lockean theory too quickly into empirically verifiable experientialism and finding himself ensconced within a firmly defined theological tradition for which no apology was needed, Edwards is therefore predisposed to focus on the trap to which reliance on experience could lead. Edwards thus betrays an awareness of the risk of naive introspection, of the possibility of deception in experience, and the fear of hypocrisy, all of which resonates well with the mind-set of Puritanism, which as a reform of the Reformation was continually asking for the testing and purifying of experience. In Proudfoot's view Edwards comes out the stronger for his skepticism, while James faces two obstacles that he is not fully able to surmount. On the one hand, James is more accepting of religious descriptive accounts, lacking the kind of skepticism that is ingrained in Edwards, while on the other hand, again linked to the aspect of description, he does not factor in the belief of the person undergoing a religious experience as constitutive of her or his religious account, seeing the account as somehow independent.

Accepting Proudfoot's analysis, I would like to emphasize slightly different points to further my interpretation of James. First, the point of skepticism in Edwards is well taken and allows us to argue for a stronger connection between Edwards and Emerson but also, indeed, James, who in my view should not so rashly be removed from theological discussion, even if he does not aim for a connection with the Puritan heritage of American culture. Here we may usefully employ the metaphor of center and margin, using a wide-angle lens to develop a broader view of the American theological landscape irrespective of ecclesial sanction. Second, I wonder whether Proudfoot does not impose an artificial criterion of a yet-to-be-developed science of religions on James by insisting that James, in his account of religious experience, acknowledge a person's belief regarding the origin of her or his experience. Since it is what provides the explanation, Proudfoot considers it constitutive of the religious experience as narrated and, consequently, sees it as a failure that James only describes the religious experience and gives his assessment. Here I want to refer back to Thomas Hardy's concept of the self-unseeing as precisely allowing James to make good use of religious narratives, not by treating them as objective or subjective first-person descriptions but by using them as indelible footprints of the "something" that happens in conversion and its sense of abiding newness.

Clearly, if I can refer back to Schleiermacher here for a moment, neither the canonical status of biblical literature nor Christology plays much of a role for James. But rather than concluding that, therefore, all of that has dropped away, is it not better to say that such aspects have all been swept up in a far more totalizing sense of religious experience? Part of the totalizing character of religious experience means that it also displays cosmic overtones; these become increasingly visible in James's later years. I see the scientist in James come to the fore in his posture as a careful reader and observer, able and willing to move the dial when needed to register what religion really means to each person and how it influences them and their lives. While Proudfoot is correct in saying that James does not consider a person's belief constitutive of their religious experience, James regards their accounts and what they have voiced in them as deeply reflective of their experience. It is those that James is very careful to read, register, and review. However much he may have wanted to be the scientist, it is the experienced and experiential voices of people like Edwards, but also as far back as Paul and Augustine, that bring James closer—I would almost say scientifically closer—to discerning their religious affections, the latter being the operative term he shares with Edwards.

Whether or not also connected to Edwards, it is certainly the cosmic that increasingly holds James's attention. Nature gives us two keys with which to test the lock. If the moral key fits the lock, and I am inclined to substitute a religious key for it here, then the universe works for that person; if not, then it does not. The fact that theological pressures of a communal or institutional nature are not an issue for James—as they are for Edwards, kindling his skepticism of experience—allows James to face a modern world in which not everybody is or needs to be religious. This means that the understanding of religion is to some extent less existential, as original sin is no longer a universal millstone around humanity's neck, and salvation is more an issue of personal self-surrender than of being taken up by a higher power. Religion in our fraught modern world cannot but have an overt anthropological outlook, and in James's *Varieties* it certainly does. But whether the loss of a meaningful institutional landscape (not yet complemented by adequate knowledge of non-Christian religions) is at issue here, or whether we should rather see James as striving for stronger, more intimate cosmic ties as our alternative home, the fact is that for James religion has cosmic aspirations. The universe is used as the polar opposite of solipsism, instilling in us and keeping alive a sense of community without constraints, an intimacy without exclusion, in sum, a plurality of equal sentiments.

Emerson: Nature as Incarnate Thought

With that, let me return to Emerson to conclude this project on "thinking nature" with some final comments on how it affects the nature of his thinking. As I stated in my introduction, Emerson's nature is, in his own words, "thoroughly mediate." In itself that comment is deceptive, as Emerson often is. This is in fact part of a longer quotation: "Nature is thoroughly mediate. It is made to serve. It receives the dominion of man as meekly as the ass on which the Saviour rode" (*Nature, CW* 1:25). Invoking the Bible as he references Christ's entry into Jerusalem on Palm Sunday, does Emerson not just restate, here in New Testament terms, the traditional stewardship model found in Genesis, and does he not thus reinforce humanity's dominion and exploitation of nature?

There are two counterarguments to this objection. The first counterargument harks back to a point I made in Chapter 1 about Emerson's sudden changes of perspective.[21] In the above passage it is man who dominates nature, with Emerson focalizing by bringing in analogies: Christ is bigger than man, and the ass is smaller than nature. In another passage, however, taken not from the book *Nature* but from the essay of the same name, we find Emerson discussing how we are totally owned by nature:

"If we consider how much we are nature's, we need not be superstitious about towns, as if that terrific or benific force did not find us there also, and fashion cities. Nature who made the mason, made the house" ("Nature," *CW* 3:106). He goes on to encourage us to "be men instead of woodchucks, and the oak and the elm shall gladly serve us." Thus, nature is prior to us and has us under its command, a command that overrides the nature/culture divide, which still keeps nature in control even as it is also "thoroughly mediate."

The second counterargument also concerns sudden shifts of emphasis, but it also concerns time, or that which is on our horizon. This point forces us to speculate about our own era as well as a future one, as Emerson moves from an apocalyptic perspective to a visionary or deep one. I want to draw out this point, therefore, by stringing together two passages that involve Emerson's take on Christianity and Christ. The first passage, from "Experience," again has an interesting oblique biblical reference, this time to Revelations and the seven seals, and betrays a loaded character:

> How easily, if fate would suffer it, we might keep forever these beautiful limits, and adjust ourselves, once for all, to the perfect calculation of the kingdom of known cause and effect. In the street and the newspapers, life appears so plain a business, that manly resolution and adherence to the multiplication table through all weathers, will insure success. But ah! presently comes a day—or is it only a half-hour, with its angel-whispering[22]—which discomfits the conclusions of nations and of years! Tomorrow again, everything looks real and angular, the habitual standards are reinstated, common sense is as rare as genius,—is the basis of genius, and experience is hands and feet to every enterprise;—and yet, he who should do his business on this understanding, would be quickly bankrupt. ("Experience," *CW* 3:39)

The passage is ominous. Emerson seems open to the possibility that the scientific, economically thriving world of steady progress could be upended, and a catastrophe of apocalyptic proportions could break in. Without wanting to insert a superficially psychological interpretation, I suggest that it is relevant that Emerson wrote this essay shortly after the death of his young son. Yet rather than see the passage in terms of personal or historical cause and effect, what I want to bring out is how Emerson insists on the need for what he calls elsewhere in the essay "the sphericity of the soul."

In the immensely capacious view that that position allows him to take, it is relevant that the second passage to which I want to point begins with the view that self-consciousness is our fall, which is also our awakening to critical thought: "It is very unhappy, but too late to be helped, the discovery we have made, that we exist. That discovery is called the Fall of Man. Ever afterwards, we suspect our instruments. We have learned that we do not see directly, but mediately ("Experience," *CW* 3:43). In continuing to speak about how our "subject-lenses," as Emerson calls them, may have a creative power but how there may also simply be no objects, which expresses his inherent skepticism, he mentions that "Nature, art, persons, letters, religions,—objects, successively tumble in and God is but one of its ideas." Grief does not help him in finding consolation, for "grief too will make us idealists" (29). Continuing to speak about humiliations and idolatries, the faults of incorrect, "non-mediate" seeing, he continues:

> People forget that it is the eye which makes the horizon, and the rounding mind's eye which makes this or that man a type or representative of humanity with the name of hero or saint. Jesus "the providential man," is a good man on whom many people are agreed that these optical laws shall take effect. By love on one part, and by forbearance to press objection on the other part, it is for a time settled, that we will look at him in the centre of the horizon, and ascribe to him the properties that will attach to any man so seen. But the longest love or aversion has a speedy term. The great and crescive self rooted in absolute nature, supplants all relative existence, and ruins the kingdom of mortal friendship and love. ("Experience," *CW* 3:44)

I have offered two counterarguments to the objection that Emerson has a traditional stewardship view of nature, which would involve human dominion over it. First, nature always trumps humanity, including humanity's dominion. Second, there is temporality to Emersonian nature, which the above two passages dealing with temporality and the deep vision arising from it illustrate.

In conclusion, what is so amazing about the last passage is that Emerson seems able to look here at the backside of the moon that is Western culture, seeing it as if from afar, and it is the sphericity of the soul that allows him to do so. This time he does not use his eye as a transparent eyeball, allowing the world to go right through him. Rather, performing another accordion-like move, he tries to actively, individually, and personally frame the culture of which he is a part by better adjusting his subject-lens. From that capacious, spherical perspective, we can see Christianity as an interlude, a time in which culture came together on this providential—that is, divine— man.[23] But just as the bond with his son was not given to Emerson for a long time, so the culture of Christianity may not be given to humankind for all of eternity; therefore, in another, perhaps more iconoclastic, but also very Emersonian reading of "providential,"[24] humanity itself needs to have the foresight of its potential vanishing. In the void that this skeptical position opens up about life as "the equator," "a narrow belt" that positions us between "lifeless science" and sensation (36), the only consolation that Emerson and we have is the affirmation that "the universe is the bride of the soul" (44), that nature and selfhood are inseparably connected.

It has been the task of the self since Eriugena, and still in Emerson, to keep that connection alive by "thinking" and "rethinking nature." They both do so in a Christian culture and for a Christian audience: Emerson co-opting Christ as the "providential" man whose divinity is needed to affirm our humanity, Eriugena including God in nature. Their originality, however, neither diminishes the religious value of their work nor puts it necessarily in contrast with the received religion of Christianity, even as it challenges ingrained structures.

With the dance of nature and self a perpetual one, the ultimate result of "thinking nature" on the nature of thinking in Emerson, and I venture to say in Eriugena as well, is not a cramped visioning of the world, as the motto of this conclusion might have one think, but something closer to the opposite: the convergence of nature with thought or, better, the release of nature into the free thought that carries us and it onward and onward: "The reality is more excellent than the report.[25] Here is no ruin, no discontinuity, no spent ball. The divine circulations never rest nor linger. Nature is the incarnation of a thought, and turns to a thought again, as ice becomes water and gas. The world is mind precipitated, and the volatile essence is forever escaping again into the state of free thought" ("Nature," *CW* 3:113).

Notes

1. On Eriugena's pantheism see Dermot Moran, *The Philosophy of John Scottus Eriugena: A Study of Idealism in the Middle Ages* (Cambridge: Cambridge University Press, 1989), 84–89; and Dermot Moran, "Pantheism from John Scottus Eriugena to Nicholas of Cusa," *American Catholic Philosophical Quarterly* 64, no. 1 (1990): 131–52. The eighteenth-century term *pantheism* itself is both problematic and protean, as is well argued by William Mander in his "Pantheism," *Stanford Encyclopedia of Philosophy Archive* (Winter 2016 Edition), https://plato.stanford.edu/archives/win2016/entries/pantheism. Recently, Mary-Jane Rubenstein has commented on the different ontologies underlying pantheism, focusing either on unity or on immanence. In the former case, oneness is ascribed to the world and God, while in the latter there is a claim of this-worldliness for the divine. See Mary-Jane Rubenstein, *Pantheologies: Gods, Worlds, Monsters* (New York: Columbia University Press, 2018), 21–22. While the circumstances of his condemnation are unclear, it would seem that Eriugena was guilty on both counts. For further comments on pantheism and my objections to the term, see Chapter 1 below.

2. The edition of the *Periphyseon* by H. J. Floss in J. P. Migne's *Patrologia Latina* 122:441A–1022C, which was standard for a long time, has now been replaced by Edouard Jeauneau's five-volume critical edition. This critical text is found in the series Corpus Christianorum Continuatio Mediaevalis [CCCM], vols. 161–65 (Turnhout: Brepols, 1996–2003). The columns of the Migne text are carried over into the critical edition. I generally quote the critical text of the *Periphyseon* according to book and Migne column numbers and have taken the English translation, and where needed adapted, from Eriugena, *Periphyseon*, trans. I. P. Sheldon-Williams, rev. J. J. O'Meara (Montreal/Washington: Bellarmin and Dumbarton Oaks, 1987).

3. Elena Lloyd-Sidle offers us an outline of the entire *Periphyseon* as an appendix to her introduction to its themes in her chapter "A Thematic Introduction to and Outline of the *Periphyseon*, for the Alumnus," in *A Companion to John Scottus Eriugena*, ed. A. Guiu (Leiden: Brill, 2020), 113–33.

4. One way of looking at pantheism is that it foregrounds an outsized notion of divine immanence (see note 1 above). It is this view that seems to have most

haunted the evaluation of the *Periphyseon*. A key text in this regard is *Periphyseon* 2.528B (here I quote from the Sheldon-Williams translation):

> M[aster]. But suppose you join the creature to the Creator so that you understand in the former nothing save Him who alone truly is—for nothing apart from Him is truly called essential since all things that are, are nothing else in so far as they are, but participation in Him who alone subsists from and through Himself—you will not deny then that Creator and creature are one?

> S[tudent]. It would not be easy for me to deny it. For it seems to me ridiculous to resist this conclusion. (127)

5. The above passage continues by emphasizing precisely nature's indivisibility, which we can even take to mean its irreducibility:

> M. So the universe, comprising God and creation, which was first divided into four forms, is reduced again to an indivisible One, being Principle as well as Cause and End.

> S. I see that we have meanwhile said enough about the universal division and unification of universal nature.

6. This passage is quoted in Branka Arsić, "Brain Walks: Emerson on Thinking," in *The Other Emerson*, ed. Branka Arsić and Cary Wolfe (Minneapolis: University of Minnesota Press, 2010), 75; also published in Branka Arsić, *On Leaving: A Reading in Emerson* (Cambridge, MA: Harvard University Press, 2010), 149.

7. The *locus classicus* of Emersonian onwardness is found in his essay "Experience" (*CW* 3:32): "The secret of the illusoriness is in the necessity of a succession of moods or objects. Gladly we would anchor, but the anchorage is quicksand. This onward trick of nature is too strong for us: *Pero si muove*. When, at night, I look at the moon and stars, I seem stationary, and they to hurry. Our love of the real draws us to permanence, but health of body consists in circulation, and sanity of mind in variety or facility of association."

8. I comment further on precisely this Emersonian citation in my conclusion.

9. Amitav Ghosh, *The Great Derangement: Climate Change and the Unthinkable* (Chicago: University of Chicago Press, 2016), 4–5. I am grateful to Charles Hallisey for this reference.

CHAPTER 1

1. Peter Dronke mentions these different conceptions of nature in the introduction to *Fabula: Explorations into the Uses of Myth in Medieval Platonism* (Leiden: Brill, 1985), 1. Referencing Eriugena, Hildegard of Bingen, Bernard Silvestris, and William of Conches, Dronke observes: "Their cosmological insights are nourished by imaginative springs as much as by the disciplined sources of abstract thinking. Theirs is a realm where sacred vision and profane myth can combine with analytic thought, poetic fantasy with physical and metaphysical speculation."

2. Spinoza, *Ethica* IVPref. and IVP4Dem. [1], Benedict de Spinoza, *Ethics*, ed.

and trans. Edwin Curley, with an introduction by Stuart Hampshire (London: Penguin, 1996), 114, 118–19.

3. Emerson uses this same distinction in his essay "Nature." See Ralph Waldo Emerson, *The Collected Works of Ralph Waldo Emerson* (hereafter *CW*), vol. 3, *Essays: Second Series*, ed. Alfred R. Ferguson and Jean Ferguson Carr (Cambridge, MA: Belknap Press of Harvard University Press, 1983), 97–114, 103–4.

4. Spinoza, *Ethica* IP29Schol, in Spinoza, *Ethics*, 20–21.

5. The debate between Karl Barth and Emil Brunner on natural theology is analyzed in Alistair McGrath, *Emil Brunner: A Reappraisal* (Oxford: Wiley Blackwell, 2014) 90–132. Barth's *Nein! Antwort an Emil Brunner* from 1934 is a radical rejection of Brunner's earlier *Natur und Gnade* from the same year. Barth rejected Brunner's natural theology as irreconcilable with God's revelation in Jesus Christ and criticized his nonexegetical approach. While both discussants largely spoke at cross-purposes, Brunner voices criticism of Barth's repeated rejection of natural theology, developed during the deteriorating political situation in Germany. Brunner did not defend natural theology but rather wanted a new natural theology, replacing Barth's *analogia fidei* with a common Christian *analogia entis*. He next developed a theological anthropology (*Der Mensch in Widerspruch*, 1937).

6. This confusion dates back to at least the seminal article by Lynn White Jr., "The Historical Roots of Our Ecological Crisis," *Science*, n.s., 155, no. 3767 (March 10, 1967): 1203–7, widely considered the clarion call for theological reflection on this matter.

7. There is an irony in juxtaposing Peter Harrison, *The Bible, Protestantism, and the Rise of Natural Science* (Cambridge: Cambridge University Press, 1998) with Hans W. Frei, *The Eclipse of Biblical Narrative: A Study in Eighteenth and Nineteenth Century Hermeneutics* (New Haven, CT: Yale University Press, 1974). Harrison considers the iconoclastic attitude of the early Protestants vis-à-vis the medieval scientific canon alongside their precision in decoding scripture determinative for the rise of early modern science, whereas Frei focuses on the decline of scriptural rootedness centuries later as linked in part to the rise of science. See also Jonathan Sheehan, *The Enlightenment Bible: Translation, Scholarship, Culture* (Princeton, NJ: Princeton University Press, 2005).

8. P. Hadot's *The Veil of Isis: An Essay on the History of the Idea of Nature* (Cambridge, MA: Belknap Press of Harvard University Press, 2008) elaborates on Heraclitus's statement that "nature loves to hide." There is no overlap with any of the thinkers discussed in the present study. Hadot's English title echoes the study by R. G. Collingwood, *The Idea of Nature* (Oxford: Oxford University Press, 1945). Similar to Hadot, Collingwood does not engage patristic or medieval thinkers, skipping from Greek cosmology to the Renaissance.

9. This echoes the opening words of Robert D. Richardson, "Emerson and Nature," in *The Cambridge Companion to Ralph Waldo Emerson*, ed. Joel Porte and Saundra Morris (Cambridge: Cambridge University Press, 1999), 97: "Explicit or

implicit in nearly everything Emerson wrote is the conviction that nature bats last, that nature is the law, the final word, the supreme court."

10. See among a plethora of available views: Robert Pogue Harrison, *Forests: The Shadow of Civilization* (Chicago: University of Chicago Press, 1992); Robert Pogue Harrison, *Gardens: An Essay on the Human Condition* (Chicago: University of Chicago Press, 2008); Robert N. Watson, *Back to Nature: The Green and the Real in the Late Renaissance* (Philadelphia: University of Pennsylvania Press, 2006); Andrea Wulf, *The Invention of Nature: Alexander von Humboldt's New World* (New York: Vintage, 2015); Bruno Latour, *The Politics of Nature: How to Bring the Sciences into Democracy* (Cambridge, MA: Harvard University Press, 2004); and Peter Harrison, *The Territories of Science and Religion* (Chicago: University of Chicago Press, 2015).

11. This is implied in Norman Wirzba, *The Paradise of God: Renewing Religion in an Ecological Age* (Oxford: Oxford University Press, 2003), 93–94: "The teachings of ecology promise a revolution in self- and cultural understanding that matches, if not exceeds, in importance the sixteenth-century Copernican astronomical revolution. Just as Copernicus forced a fundamental reorientation in how the universe and thus also we ourselves are to be perceived and understood, so too ecological insight compels a transformation in basic presuppositions about nature and human nature."

12. Thrownness represents Heidegger's *Geworfenheit*, the state of being thrown into the world. See Martin Heidegger, *Being and Time: A Translation of Sein und Zeit*, trans. Joan Stambaugh, rev. and with a foreword by Dennis J. Schmidt (Albany: State University of New York Press, 2010), 127: "We shall call this character of being of Da-sein which is veiled in its whence and whither, but in itself all the more disclosed, the 'this that it is,' the *thrownness* of its being into its there; it is thrown in such a way that it is the there of being-in-the-world. The expression thrownness is meant to suggest the *facticity of its being delivered over*."

13. It is a unique feature of (both the Hebrew and) the Christian conception of creation that God creates exclusively by speaking, with John 1 a deliberate echo of Genesis 1.

14. This was indeed the reason why Brunner went from creation to theological anthropology; see note 5 above.

15. My course here runs counter to Norman Wirzba, *From Nature to Creation: A Christian Vision for Understanding and Loving Our World* (Grand Rapids, MI: Baker Academic, 2015).

16. See Paul DeHart, "f(S) I/s: The Instance of Pattern, or Kathryn Tanner's Trinitarianism," in *The Gift of Theology: The Contribution of Kathryn Tanner*, ed. Rosemary P. Carbine and Hilda P. Koster (Minneapolis, MN: Fortress, 2015), 29–55, 31: "Ever since her first book, *God and Creation*, Tanner's work has rightly been associated with an acute sensitivity to the conceptually unique causal complexity involved whenever divine activity is brought into relation with creaturely events. Her often-invoked 'non-competitive' account of relationality refers to her justified (and quite traditional) strictures against allowing divine causality to parallel created

causality, as if they were different instances of shared agency." See also Eugene F. Rogers Jr., "Tanner's Non-competitive Account and the Blood of Christ: Where Eucharistic Theology Meets the Evolution of Ritual," in Carbine and Koster, *The Gift of Theology*, 139–57.

17. See Kathryn Tanner, *God and Creation in Christian Theology: Tyranny or Empowerment?* (Minneapolis, MN: Fortress, 2005), 79: "God's transcendence over and against the world and God's immanent presence within it become non-exclusive possibilities, then, when God's primary transcendence is a self-determined transcendence essentially independent of a contrast with the non-divine. The ordinarily mutually exclusive predicates 'transcendence' and 'immanence' may both be applied to one and the same self-determining divine subject. The apparent mutual exclusiveness of those terms has to give way before a God who is freely self-determining." This matters because, as affirmed in her essay "Human Freedom, Human Sin, and God the Creator," in *The God Who Acts: Philosophical and Theological Explorations*, ed. Thomas F. Tracy (University Park: Pennsylvania State University Press, 2005), 112, God's role as creator is a lynchpin of Christian identity.

18. See W. Otten, *From Paradise to Paradigm: A Study of Twelfth-Century Humanism* (Leiden: Brill, 2004), 215–55 (on Bernard) and 256–85 (on Alan). See also my essay "Medieval Latin Humanism," in *Encyclopedia of Mediterranean Humanism*, ed. Houari Touati, Spring 2016, www.encyclopedie-humanisme .com/?Medieval-Latin-Humanism.

19. See G. D. Economou, *The Goddess Natura in Medieval Literature* (Notre Dame, IN: Notre Dame University Press, 2002), 28–52.

20. Prior to outright criticism there was the budding self-censorship of twelfth-century poetic works. See Peter Godman, *The Silent Masters: Latin Literature and Its Censors in the High Middle Ages* (Princeton, NJ: Princeton University Press, 2001).

21. See W. Otten, "Creation and Epiphanic Incarnation: Reflections on the Future of Natural Theology from an Eriugenian-Emersonian Perspective," in *On Religion and Memory*, ed. B. S. Hellemans, W. Otten, and M. B. Pranger (New York: Fordham University Press, 2013), 64–88, esp. 69–74.

22. In other respects Tanner's creation has remarkable Thomistic overtones, as she dispenses with temporality, one of this book's themes. See her "Human Freedom, Human Sin," 113: "One can say . . . that God's creation of the world loses any specific reference to a beginning time or initiating moment: to be created is to be in a relation of dependence upon God that holds whenever and for however long one exists."

23. For background on the "Pantheismusstreit," see Gérard Vallée, *The Spinoza Conversations between Lessing and Jacobi: Texts with Excerpts from the Ensuing Controversy*, introd. Gérard Vallée, trans. G. Vallée, J. B. Lawson, and C. G. Chapple (Lanham, MD: University Press of America, 1988). In 1783, through the mediation of Elise Reimarus, Lessing was outed to Moses Mendelssohn about his Spinozist sympathies by Jacobi. Mendelssohn saw in Lessing a representative

of a "refined" Spinozism, or rather a pantheism that might in fact be closer to panentheism. As such, "panentheism postulates a dipolar nature of the divinity according to which God, through his relative aspect, includes the world and, through his absolute aspect, is distinct from the world" (47). Whether the charge that Lessing in his later days was a Spinozist is true remains not so much unclear as unelucidated.

24. See Terrence W. Tilley, *Evils of Theodicy* (Eugene, OR: Wipf and Stock, 2000), 221–55, on the Enlightenment construction and employment of theodicy since Leibniz's use of it as a title in 1710. It is important to note that theodicy is inseparable from the pantheism debate, for "once determinism is espoused, a theodicy is needed" (Vallée, *The Spinoza Conversations*, 15).

25. See Julia A. Lamm, *The Living God: Schleiermacher's Theological Appropriation of Spinoza* (University Park: Pennsylvania State University Press, 1996), 95–126. Lamm deems Schleiermacher's Explanations to the third edition of the *Reden* and the second edition of *Der christliche Glaube* (*The Christian Faith*) pantheistic in two qualified ways—namely, in Schleiermacher's tolerance/espousal of a nonpersonal conception of God and in his adaptation of the dictum *One and All* (also used by Lessing, *Hen Kai Pan*). Yet she regards his Spinozism as broader than his pantheism (96). See further below Chapter 4.

26. The thesis of my *From Paradise to Paradigm* is that twelfth-century naturalizing literature and thought collapsed under the weight of an increasingly complex set of self-referential resonances and are therefore not direct run-ups to thirteenth-century natural philosophy arising from the influx of Aristotelianism.

27. On *integumentum*, see E. Jeauneau, "L'usage de la notion d'*integumentum* à travers les gloses de Guillaume de Conches," *Archives d'histoire doctrinale et littéraire du Moyen Age*, vol. 24 (1957): 35–100; repr. in E. Jeauneau, *Lectio philosophorum: Recherches sur l'école de Chartres* (Amsterdam: Hakkert, 1973), 127–92. See also Dronke, *Fabula*, 25–32, 52.

28. Only two of the three projected volumes appeared: R. W. Southern, *Scholastic Humanism and the Unification of Europe*, vol. 1, *Foundations* (Oxford: Blackwell, 1995); and vol. 2, *The Heroic Age* (Oxford: Blackwell, 2001).

29. This fissure is prepared in the twelfth century. For my view of the dissolution and disappearance of twelfth-century nature speculation aside from its vigorous afterlife in vernacular poetry, see W. Otten, "Nature and the Representation of Divine Creation in the Twelfth Century," in *Communities of Learning: Networks and the Shaping of Intellectual Identity in Europe, 1100–1500*, ed. C. Mews and J. N. Crossley (Turnhout: Brepols, 2011), 57–74. The counterview that scholastic *philosophia naturalis* is a continuation of twelfth-century natural thought is found in A. Speer, *Die entdeckte Natur: Untersuchungen zu Begründungsversuchen einer "scientia naturalis" im 12. Jahrhundert* (Leiden: Brill, 1995).

30. See Otten, *From Paradise to Paradigm*, 83–128.

31. On the twelfth-century allegorists see Barbara Newman, *God and the*

Goddesses: Vision, Poetry, and Belief in the Middle Ages (Philadelphia: University of Pennsylvania Press, 2003), 51–137; and, more recently, Kellie Robertson, *Nature Speaks: Medieval Literature and Aristotelian Philosophy* (Philadelphia: University of Pennsylvania Press, 2017). On Bonaventure see Christopher M. Cullen, *Bonaventure* (Oxford: Oxford University Press, 2006); and Denys Turner, *The Darkness of God: Negativity in Christian Mysticism* (Cambridge: Cambridge University Press, 1995), 102–34 (on the path of humanity's return through creation to God).

32. See on this W. Otten, "Suspended between Cosmology and Anthropology: *Natura*'s Bond in Eriugena's *Periphyseon*," in *A Companion to John Scottus Eriugena*, ed. Adrian N. Guiu (Leiden: Brill, 2020), 189–212.

33. For earlier comparisons between them see W. Otten, "Eriugena and Emerson on Nature and the Self," in *Eriugena and Creation*, ed. M. I. Allen and W. Otten (Turnhout: Brepols, 2014), 503–38; and Otten, "Creation and Epiphanic Incarnation." See also my "Nature as Religious Force in Eriugena and Emerson," in *Religion: Beyond a Concept*, ed. Hent de Vries (New York: Fordham University Press, 2008), 354–67.

34. The reversal of Freud's *Civilization and Its Discontents* is my organizing principle for the article "Christianity's Content: (Neo)Platonism in the Middle Ages, Its Theoretical and Theological Appeal," *Numen* 63, no. 2–3 (2016): 245–70.

35. See especially Stanley Cavell, *The Claim of Reason: Wittgenstein, Skepticism, Morality, and Tragedy* (Oxford: Oxford University Press, 1979).

36. Cavell analyzes Emerson's response to Descartes's *cogito* as found in the essay "Self-Reliance" in his essay "Being Odd, Getting Even (Descartes, Emerson, Poe)," in *In Quest of the Ordinary: Lines of Skepticism and Romanticism* (Chicago: University of Chicago Press, 1994), 105–30; repr. in Stanley Cavell, *Emerson's Transcendental Etudes* (Stanford: Stanford University Press, 2003), 83–109. For Emerson's response to Kant see Stanley Cavell, "Emerson, Coleridge, Kant (Terms as Conditions)," in *In Quest of the Ordinary*, 27–49; repr. in Cavell, *Emerson's Transcendental Etudes*, 59–82.

37. Cavell references Thoreau's *Walden* as emblematic in this regard; see his *The Senses of Walden: An Expanded Edition* (Chicago: University of Chicago Press, 1992), xiii–xv. On philosophy as a mix of philosophy and literature see his essay "Thinking of Emerson," in *The Senses of Walden*, 129–35; repr. in *Emerson's Transcendental Etudes*, 10–32. I have highlighted the prephilosophical moment, the philosophizing rather than the philosophical product, with the term *conversation* (see Stanley Cavell, *Conditions Handsome and Unhandsome: The Constitution of Emersonian Perfectionism* [Chicago: University of Chicago Press, 1990], xx; and Otten, "Nature as Religious Force," 362–66) in part to signal a convergence with David Tracy's description of the "classic" as eliciting perennial conversation. See David Tracy, *Plurality and Ambiguity: Hermeneutics, Religion, Hope* (Chicago: University of Chicago Press, 1987), 20: "Conversation in its primary form is an exploration of possibilities in the search for truth."

38. Cavell, *Emerson's Transcendental Etudes*, 16, 38–40.

39. Two important examples are Lawrence Buell, *The Environmental Imagination: Thoreau, Nature Writing, and the Formation of American Culture* (Cambridge: Belknap Press of Harvard University Press, 1995); and Catherine L. Albanese, *A Republic of Mind and Spirit: A Cultural History of American Metaphysical Religion* (New Haven, CT: Yale University Press, 2007).

40. See Lawrence Buell, *Emerson* (Cambridge, MA: Belknap Press of Harvard University Press, 2003); Branka Arsić, *On Leaving: A Reading in Emerson* (Cambridge, MA: Harvard University Press, 2010); and Branka Arsić and Cary Wolfe, eds., *The Other Emerson* (Minneapolis: University of Minnesota Press, 2010).

41. I have tried to do this in W. Otten, "Eriugena, Emerson, and the Poetics of Universal Nature," in *Metaphysical Patterns in Platonism: Ancient, Medieval, Renaissance, and Modern Times*, ed. R. Berchman and J. Finamore (New Orleans: University Press of the South, 2007), 147–63. Reprinted by the Prometheus Trust: Wiltshire, UK, 2014, 123–37.

42. The text of *Nature* and the "Divinity School Address" is found in Ralph Waldo Emerson, *CW*, vol. 1, *Nature, Addresses, and Lectures* (Cambridge, MA: Belknap Press of Harvard University Press, 1971), 7–45 and 76–93, respectively. For "Circles" see *CW*, vol. 2, *Essays: First Series* (Cambridge, MA: Belknap Press of Harvard University Press, 1979), 177–90.

43. See Eduardo Cadava, *Emerson and the Climates of History* (Stanford: Stanford University Press, 1997), 97–106.

44. Interestingly, Emerson actually speaks of both dogmas and prayers, presumably formulaic liturgical prayers: "The prayers and even the dogmas of our church, are like the zodiac of Denderah, and the astronomical monuments of the Hindoos, wholly insulated from anything now extant in the life and business of the people" (*CW* 1:86).

45. On Schleiermacher see John Dillenberger and Claude Welch, *Protestant Christianity Interpreted through Its Development*, 2nd ed. (Upper Saddle River, NJ: Prentice Hall, 1998), 163–69. See also Friedrich Schleiermacher, *Der christliche Glaube nach den Grundsätzen der evangelischen Kirche im Zusammenhange dargestellt*, zweite Auflage (1830/31), §§15–16, ed. Rolf Schäfer (Berlin: De Gruyter, 2008), esp. *Kritische Gesamtausgabe* (*KGA*) 1/13.1:127: "Christliche Glaubenssätze sind Auffassungen der christlich frommen Gemüthszustände in der Rede dargestellt." I address Schleiermacher's position further in Chapter 4, as well as in my conclusion.

46. On the point of stabilizing progress see John Henry Newman's 1845 text, *An Essay on the Development of Christian Doctrine* (Cambridge: Cambridge University Press, 2010). On Newman's influence, through Yves Congar, on the Second Vatican Council, see Andrew Meszaros, *The Prophetic Church: History and Doctrinal Development in John Henry Newman and Yves Congar* (Oxford: Oxford University Press, 2016), 60–95.

47. It seems Emerson uses *homonymy* or close verbal association as a privileged form of thinking onward. In the "Divinity School Address," he productively associates seer and sayer, as the prophet embodies both: "Always the seer is a sayer" (*CW* 1:84). Here I want to point to the connection between the bread of life and the breath of new life.

48. See F. W. Nietzsche, *The Complete Works of Friedrich Nietzsche*, vol. 8, *Beyond Good and Evil / On the Genealogy of Morals*, trans. Adrian Del Caro (Stanford: Stanford University Press, 2014). The importance of history for Nietzsche is also evident from his *On the Advantage and Disadvantage of History for Life*, trans. Peter Preuss (Indianapolis, IN: Hackett, 1980).

49. Given my focus on religion, I have highlighted Schleiermacher and Newman to offset Nietzsche's more complex genealogical view of history. The tension between Nietzsche and his senior colleague in history at Basel, Jakob Burkhardt, further reveals history to be the dominant factor behind nineteenth-century cultural views. See Emil W. Busch, *Burckhardt und Nietzsche im Revolutionszeitalter* (München: Fink, 2012). Emerson's position hence appears to run counter to the historicizing mind-set of the age.

50. "From the views I have already expressed, you will infer the sad conviction, which I share, I believe, with numbers, of the universal decay and now almost death of faith in society. The soul is not preached. The Church seems to totter to its fall, almost all life extinct" ("Address," *CW* 1:84). See also "We have contrasted the Church with the Soul" ("Address," *CW* 1:89).

51. See Cavell, *Emerson's Transcendental Etudes*, 6.

52. Cavell, *Emerson's Transcendental Etudes*, 25 ("An Emerson Mood"; repr. from *The Senses of Walden*, 149–50), speaks in this context of Emerson's "physiognomy of the ordinary, a form of what Kierkegaard calls the perception of the sublime in the everyday."

53. Interestingly, Emerson casts this utterance of Jesus as a jubilee of sublime emotion.

54. The medieval tradition also has examples of disassociating the miraculous from the monstrous. An example of a normalizing miracle is the action whereby St. Benedict makes the earth accept the body of a disobedient young monk, which it initially spit out (Gregory the Great, "Life of Benedict," *The Dialogues* bk. 2, chap. 24). See on this W. Otten, "Ideals of Community in Late Antiquity: John Cassian and Gregory the Great on Communicating Sanctity," in *Seeing the Invisible in Late Antiquity and the Early Middle Ages: Papers from "Verbal and Pictorial Imaging: Representing and Accessing Experience of the Invisible: 400–1000,"* ed. G. de Nie, K. F. Morrison, H. L. Kessler, and M. Mostert (Turnhout: Brepols, 2005), 130–34.

55. See "Experience," in *CW* 3:30: "Life is a train of moods like a string of beads, and, as we pass through them, they prove to be many-colored lenses which paint

the world their own hue, and each shows only what lies in its focus." On Emerson as a philosopher of moods see also Cavell, *Emerson's Transcendental Etudes*, 26–28 ("An Emerson Mood"); and Arsić, *On Leaving*, 135–40.

56. Note that in *Nature* Emerson states: "Nature never became a toy to a wise spirit" (*CW* 1:9).

57. Emerson describes it as follows: "the Sabbath, the jubilee of the whole world; whose light dawns welcome alike into the closet of the philosopher, into the garret of toil, and into prison cells, and everywhere suggests, even to the vile, a thought of the dignity of spiritual being" (*CW* 1:92).

58. See Emerson, "Self-Reliance," *CW* 2:37; see also Cavell, *Emerson's Transcendental Etudes*, 93.

59. See W. Otten, "Nature as a Theological Problem: An Emersonian Response to Lynn White," in *Responsibility and the Enhancement of Life: Essays in Honor of William Schweiker*, ed. Günter Thomas and Heike Springhart (Leipzig: Evangelische Verlagsanstalt, 2017), 265–80.

60. I take this to be a valid criticism that Bruno Latour makes of the bicameral view of nature (political versus scientific, but also natural versus cultural) from which he wants to bring us around to a new and collective *cosmos*. See Otten, "Creation and Epiphanic Incarnation," 67–69, with reference to Latour, *Politics of Nature*, 4–5, 9–52.

61. On this final coincidence of nature and self see Emerson, "The American Scholar": "And, in fine, the ancient precept 'Know Thyself,' and the modern precept, 'Study nature,' become at last one maxim" (*CW* 1:55).

62. While Emerson takes this quote from sources that attribute it to Augustine (see apparatus, *CW* 2:253–54), the earliest occurrence of the infinite sphere is in the *Book of the XXIV Philosophers*, definition no. 2: "Deus est sphaera infinita cuius centrum est ubique, circumferentia nusquam." The work, which contains twenty-four commented propositions about God, has most recently been attributed to Marius Victorinus, although the earliest manuscript stems from twelfth-century Laon. On Alan of Lille's twelfth-century "Discourse on the Intelligible Sphere" (*Sermo de sphaera intelligibili*) see Dronke, *Fabula*, 144–53.

63. Emersonian perfectionism for Cavell does not mean perfectibility; rather, it signals that for Emerson "each state of the self is, so to speak, final: each state constitutes a world (a circle, Emerson says), and it is one each one also desires (barring inner or outer catastrophes)." See Cavell, *Conditions Handsome and Unhandsome*, 3.

64. See Jean-Luc Marion, "A Saturated Phenomenon," *Filozofia* 62, no. 5 (2007): 378–402.

65. The title should indeed read *Periphyseon* (Peri Physeōn = On Natures). The title *De divisione naturae* (On the division of nature) obscures the original and probably intentional echo of the title of Origen's *On First Principles* (Peri Archōn), a point first observed by Henri de Lubac. See E. Jeauneau, "From Origen's *Periarchon*

to Eriugena's *Periphyseon*," in *Eriugena and Creation*, ed. W. Otten and M. I. Allen (Turnhout: Brepols, 2014), 139–40.

66. See John J. O'Meara, *Eriugena* (Oxford: Clarendon, 1988), 216–17.

67. See note 30 above.

68. See Emerson in *Nature*: "All things with which we deal, preach to us. What is a farm but a mute gospel?" (*CW* 1:26).

69. See *Periphyseon* 1, CCCM 161:3, lines 1–2.

70. From the end of the work's first paragraph—"For nothing at all can occur in our thoughts that could fall outside this name" (that is, *natura*)—the overwhelming size of *natura* overshadows the role of the pondering self, even if the self continues to insert itself in its unfolding.

71. See Eriugena, *Periphyseon* 1.441B–42A.

72. For the Irish influence on Eriugena's thought and writing, see Alfred K. Siewers, *Strange Beauty: Ecocritical Approaches to Early Medieval Landscape* (New York: Palgrave MacMillan, 2009), 67–84, commenting on the importance of the sea, paradise, and the fourfold.

73. See E. Jeauneau, "Le symbolisme de la mer chez Jean Scot Érigène," in E. Jeauneau, *Études Érigéniennes* (Paris: Études Augustiniennes, 1987), 289–96. See also his "L'effort, le labeur," in *Quatre thèmes érigéniens* (Montreal: Vrin, 1978), 47–59; repr. in *Études Érigéniennes*, 243–55.

74. In fact, Eriugena's turn to hexaemeral exegesis starts earlier, in book 3.690C; Sheldon-Williams transl., 319. I comment further on the *hexaemeron* tradition in Chapter 3.

75. See Reinhold R. Grimm, *Paradisus coelestis, paradisus terrestris: Zur Auslegungsgeschichte des Paradieses im Abendland bis um 1200* (Munich: Fink, 1977).

76. See W. Otten, "The Long Shadow of Human Sin: Augustine on Adam, Eve, and the Fall," in *Out of Paradise: Eve and Adam and Their Interpreters*, ed. B. E. J. H. Becking and S. A. Hennecke (Sheffield: Sheffield Phoenix Press, 2010), 29–49.

77. I have described the *Periphyseon*'s double journey in W. Otten, "Eriugena on Natures (Created, Human and Divine): From Christian-Platonic Metaphysics to Early-Medieval Protreptic," in *Philosophie et théologie chez Jean Scot Érigène*, ed. Isabelle Moulin (Paris: Vrin, 2016), 113–33.

78. For my criticism of scholasticism as the standard academic theological discourse, see W. Otten, "On *Sacred Attunement*, Its Meaning and Consequences: A Meditation on Christian Theology," *Journal of Religion* 93, no. 4 (2013): 478–94.

79. See Edward Cuthbert Butler, *Western Mysticism: The Teaching of Saint Augustine, Gregory and Bernard on Contemplation and the Contemplative Life* (New York: Dover, 2003). This book is central in the debate between Denys Turner and Bernard McGinn on mysticism. See also W. Otten, "Le langage de l'union mystique: Le désir et le corps dans l'œuvre de Jean Scot Érigène et de Maître Eckhart," *Les études philosophiques*, no. 1 (Jan. 2013): 121–24.

80. For the pressures put on Anselm's so-called ontological argument, see W.

Otten, "Religion as *Exercitatio Mentis*: A Case for Theology as a Humanist Discipline," in *Christian Humanism: Essays Offered to Arjo Vanderjagt on the Occasion of his Sixtieth Birthday*, ed. Z. R. W. M. von Martels and A. MacDonald (Leiden: Brill, 2009), 67–69.

81. See B. McGinn, *The Growth of Mysticism: Gregory the Great through the 12th Century*, vol. 2 of *The Presence of God* (New York: Crossroad, 1994), 80–118. McGinn sees as the heart of Eriugena's mysticism the idea of "unfulfilled fulfillment" (118).

82. See B. McGinn, *The Foundations of Mysticism: Origins to the Fifth Century*, vol. 1 of *The Presence of God* (New York: Crossroad, 1992), xvii: "Thus we can say that the mystical element in Christianity is that part of its belief and practices that concerns the preparation for, the consciousness of, and the reaction to what can be described as the immediate or direct presence of God."

83. Herein precisely I see the contrast between Eriugena and Eckhart. See Otten, "Le langage de l'union mystique."

84. See Dermot Moran, "*Spiritualis incrassatio*: Eriugena's Intellectualist Immaterialism: Is It an Idealism?" in *Eriugena, Berkeley, and the Idealist Tradition*, ed. Stephen Gersh and Dermot Moran (Notre Dame, IN: Notre Dame Press, 2006), 123–50. See also Dermot Moran, "Jean Scot Érigène, la connaissance de soi et la tradition idéaliste," *Les études philosophiques*, no. 1 (Jan. 2013): 29–56.

85. Dermot Moran, *The Philosophy of John Scottus Eriugena: A Study of Idealism in the Middle Ages* (Cambridge: Cambridge University Press, 1989), 282.

86. See Dermot Moran, "Idealism in Medieval Philosophy: The Case of Johannes Scottus Eriugena," *Medieval Philosophy and Theology* 8, no. 1 (1999): 53–82. Cf. Myles Burnyeat, "Idealism and Greek Philosophy: What Descartes Saw and Berkeley Missed," *Philosophical Review* 91, no. 1 (1982): 3–40.

87. Moran, "Idealism in Medieval Philosophy," 54–55. For an earlier problematization of this view see W. Otten, "Realized Eschatology or Philosophical Idealism: The Case of Eriugena's *Periphyseon*," in *Ende und Vollendung: Eschatologische Perspektiven im Mittelalter*, ed. J. A. Aertsen and M. Pickavé (New York: De Gruyter, 2002), 373–87.

88. Moran, "*Spiritualis incrassatio*," 124.

89. Eriugena uses the expression *imago imaginis* in *Periphyseon* 4.790A and 4.791A, deriving it from Gregory of Nyssa's *On the Making of Man*, which he had translated as *Sermo De Imagine*. He employs it further when talking about the return of the body to the soul in 5.952C:

> The transmutations of our bodies not only into spiritual bodies but also into souls comes about as the result of natural necessity, so that, just as the rational soul, which is created in the Image of God, shall return into Him of Whom it is the image and likeness, so too the body, which is created in the image of the soul and is, as it were, an image of an image, shall, laying aside the earthly weight of its corporeality, return

into its cause, which is the soul, and through it as through an intermediary into God Himself, Who is the One Cause of all things. (Sheldon-Williams transl., 632)

I address the question of whether we need to understand return only as a metaphysical stage in my Postscript below.

90. In *Periphyseon* 4.797D–99A Eriugena cites Gregory's *Sermo 'De imagine'* 18 (*De hominis opificio* 17). See 4.797D: "Therefore the life of those who have been restored to that which is properly held to be the life of the angels, was itself before the Fall an angelic life" (Sheldon-Williams transl., 448). Following Gregory, he speculates that humans had an angelic mode of procreation before being outfitted with a corporeal, gendered body that made them dependent on sexual procreation after the Fall. This superadded body will be stripped at the end of time.

91. J. Warren Smith, *Passion and Paradise: Human and Divine Emotion in the Thought of Gregory of Nyssa* (New York: Herder-Crossroad, 2004), 33–47, recognizes the tensions in Gregory's account of creation, reflecting in part the influence of Philo, but does not take Gregory to espouse the view of a double creation. In Eriugena's reading, Gregory does subscribe to a double creation.

92. Here as elsewhere Eriugena follows Gregory, but once embedded in the *Periphyseon*'s narrative, Gregory's position takes on a different meaning. See, e.g., W. Otten, *The Anthropology of Johannes Scottus Eriugena* (Leiden: Brill, 1991), 190–219, on the priority of sinful created nature over humanity's ontologically higher status as a primordial cause in Eriugena. See on this also the Postscript below.

93. Eriugena cites Augustine's *De ciuitate* 20:14 twice, in 5.866D–67A and 5.992D–93A. He states in the latter passage: "we, following in his [i.e., Augustine's] and others' footsteps, [say] that everything in this world which is sensible and extended in space and time, and everything that is susceptible to change will perish, that is to say, will pass into its substance or nature, but that its nature, which is contained immutably and indestructibly after an incorporeal and intelligible model in its Primordial Causes, will endure forever" (Sheldon-Williams transl., 679).

94. For a longer excursion on this problem see W. Otten, "Does the Canon Need Converting? A Meditation on Augustine's *Soliloquies*, Eriugena's *Periphyseon*, and the Dialogue with the Religious Past," in *How the West Was Won: Essays on Literary Imagination, the Canon, and the Christian Middle Ages for Burcht Pranger*, ed. W. Otten, A. Vanderjagt, and H. de Vries (Leiden: Brill, 2010), 195–223.

95. To seal this claim, we have to deal with the return, which in *Periphyseon* 5 goes from matter to spirit. For a further discussion, with relevant passages, see W. Otten, "The Dialectic of the Return in Eriugena's *Periphyseon*," *Harvard Theological Review* 84, no. 4 (1991): 412–21.

96. *Periphyseon* 1.449A–B; Sheldon-Williams transl., 34: "For he [scil. Maximus] says that theophany is effected from no other [cause] but God, but that it happens as a result of the condescension of the Divine Word, that is, of the only begotten Son Who is the Wisdom of the Father, downwards as it were, upon human nature

which was created and purified by Him, and of the exaltation upwards of human nature to the aforesaid Word by divine love."

97.   On the important relational roles of Master and Student in the dialogical *Periphyseon* see Elizabeth Kendig, "La forme dialogique dans le *Periphyseon*: Recréer l'esprit," *Les études philosophiques*, no. 1 (Jan. 2013): 101–19.

98.   On Emerson's "religious radicalisms" see Buell, *Emerson*, 158–98.

99.   See on this George J. Stack, *Nietzsche and Emerson: An Elective Affinity* (Athens: Ohio University Press, 1992), 1–70.

100.   See Henry Stanley Haskins, *Meditations in Wall Street*, with an introduction by Albert Jay Nock (New York: William Morrow, 1940), 131. The book was originally published anonymously and was much later discovered to be by Haskins. On Henry Stanley Haskins see http://quoteinvestigator.com/2011/01/11/what-lies-within.

101.   See Emerson, "Self-Reliance," *CW* 2:30. See also Cavell, "An Emerson Mood," in *The Senses of Walden*, 152–60.

102.   See Cavell, *The Senses of Walden*, 30, on the relation between counting and accounting in Thoreau: "This is what those lists of numbers, calibrated to the half cent, mean in *Walden*. They of course are parodies of America's methods of evaluation; and they are emblems of what the writer wants from writing, as he keeps insisting in calling his book an account."

103.   Emerson, "Experience," *CW* 3:48. See also Robert D. Richardson Jr., *Emerson: The Mind on Fire* (Berkeley: University of California Press, 1995), 400–403.

CHAPTER 2

1.   I use *cosmos* and *creation* in this chapter interchangeably as the totality of reality evoked by Maximus, who lacks Eriugena's divisional setup but for whom God, and certainly Christ, acts through and by means of nature. In Maximus God is not an explicit part of nature or the universe, but there is a strong Christological coloration of cosmic reality. Thus, the Christological cosmos is the template of Maximian salvation. Also, while there is an absence of selfhood in Maximus, the Adamic paradigm is prevalent. Later in the chapter, I follow Maximus in talking about the parallelism of creation and scripture, alternating creation with Eriugena's parlance of scripture and nature.

2.   The Origenian position to which I refer, and which will also feature in Chapter 3, is that God created a spiritual reality before creating a material or corporeal one. The latter was a divine correction to preserve humanity's possibility of attaining the *pleroma*, the eschatological fullness at the end of time. For a succinct exposition see Henry Chadwick, "Origen," in *The Cambridge History of Later Greek and Early Medieval Philosophy*, ed. A. H. Armstrong (Cambridge: Cambridge University Press, 1967), 182–92. See also Peter Brown, *The Body and Society: Men, Women,*

and Sexual Renunciation in Early Christianity (New York: Columbia University Press, 1988), 160–77.

3. Liturgy in Maximus is rooted in but not thereby identical with ascetic practice. While I do not thematize this differentiation here, I do at the end of this chapter and return to it in the book's Conclusion.

4. For this chapter's analysis I draw on two recent articles: W. Otten, "West and East: Prayer and Cosmos in Augustine and Maximus Confessor," in *The Early Christian Mystagogy of Prayer*, ed. P. van Geest, H. van Loon, and G. de Nie (Leuven: Peeters, 2019), 319–37; and, especially, "Cosmos and Liturgy from Maximus to Hans Urs von Balthasar (with an excursion on H. J. Schulz)," in *Sanctifying Texts, Transforming Rituals: Encounters in Liturgical Studies*, ed. P. van Geest, M. Poorthuis, and H. E. G. Rose (Leiden, Brill, 2017), 153–69.

5. See Hans Urs von Balthasar, *Cosmic Liturgy: The Universe according to Maximus the Confessor*, trans. Brian E. Daley (San Francisco: Ignatius Press, 2003).

6. It takes some detective work to go through the various positions of Schulz, whose work I first read in the late 1970s. The original version I then read and have used here is Hans-Joachim Schulz, *Die byzantinische Liturgie: Vom Werden ihrer Symbolgestalt* (Freiburg: Lambertus, 1964), 81–90. I have consulted but not relied on the English translation by M. J. O'Connell, *The Byzantine Liturgy: Symbolic Structure and Faith Expression* (New York: Pueblo, 1986), 43–49, which is based on a 1980 edition of the same work. I have also consulted the updated version: H. J. Schulz, *Die byzantinische Liturgie: Glaubenszeugnis und Symbolgestalt* (Trier: Paulinus, 2000), 117–24, in which some of the objectionable passages I cite in the text have been removed. But Schulz's adjusted interpretation is unsatisfactory in another way insofar as he treats Maximus now as a mystical-ascetic rather than an ecclesial-liturgical thinker. This underscores my point in Chapter 1 that the category of mysticism is used to absorb thought, in this case Maximian liturgical ideas, that is theologically or ecclesially suspect. In the 2000 edition Schulz dates the *Mystagogy* to Maximus's monastic period before 626 CE.

7. See Pauline Allen and Bronwen Neil, eds., *The Oxford Handbook of Maximus the Confessor* (Oxford: Oxford University Press, 2015).

8. See on this, for example, Stephen Gersh, "*Per se ipsum*: The Problem of Immediate and Mediate Causation in Eriugena and His Neoplatonic Predecessors," in *Jean Scot Érigène et l'histoire de la philosophie: Laon 7–12 juillet 1975*, ed. R. Roques (Paris: CNRS, 1977), 367–76.

9. See Henri de Lubac, *History and Spirit: The Understanding of Scripture according to Origen* (San Francisco: Ignatius, 2007).

10. For Chalcedon's *definitio fidei* see Norman P. Tanner, S.J., *Decrees of the Ecumenical Councils*, vol. 1, *Nicaea I to Lateran V* (London/Washington: Sheed and Ward/Georgetown University Press, 1990), 86–88:

So, following the saintly fathers, we will with one voice teach the confession of one
and the same Son, our Lord Jesus Christ: the same perfect in divinity and perfect
in humanity, the same truly God and truly man, of a rational soul and a body; consubstantial with the Father as regards his divinity, and the same consubstantial with
us as regards his humanity; like us in all respects except for sin: begotten before the
ages from the Father as regards his divinity, and in the last days the same for us and
for our salvation from Mary, the virgin God-bearer, as regards his humanity; one
and the same Christ, Son, Lord, only-begotten, acknowledged in two natures which
undergo no confusion, no change, no division, no separation; at no point was the
difference between the natures taken away through the union, but rather the property of both natures is preserved and comes together into a single person and a single
subsistent being; he is not parted or divided into two persons, but is one and the
same only-begotten Son, God, Word, Lord Jesus Christ, just as the prophets taught
from the beginning about him, and as the Lord Jesus Christ himself instructed us,
and as the creed of the fathers handed it down to us.

11.   The translation of the *Mystagogy* is taken from George C. Berthold, trans.,
Maximus Confessor: Selected Writings (Mahwah, NJ: Paulist Press, 1985), 181–225.
For chap. 21, see p. 203.

12.   Berthold, *Maximus Confessor, Mystagogy* 21, 203. For the Greek text of the
omitted part see Christian Boudignon, ed., *Maximi Confessoris Mystagogia: Una
cum latina interpretatione Anastasii Bibliothecarii* (Turnhout: Brepols, 2001), 48,
lines 768–72.

13.   In the updated 2000 edition Schulz has omitted this analysis and inserted
an explicit comment to Maximus's referencing of the Eucharist in *Mystagogy* 24.

14.   Schulz, *Die byzantinische Liturgie*, 89. Maximus, *Maximi Confessoris Mystagogia*, 69:58, lines 930–35. Berthold, *Maximus Confessor, Mystagogy* 24, 207.

15.   Schulz, *Die byzantinische Liturgie*, 90: "Für die Deutung der liturgischen
Symbolik freilich hat das dyadische Schema des Maximos oft ein Übergehen jener
mittleren Symbolbedeutungen zur Folge, die in Anwendung des dreigliederigen Be-
griffsschemas der scholastischen Sakramententheologie als 'res et sacramentum' zu
bezeichnen wären." This phrase was still in the 1980 edition, on which the English
translation is based, but is no longer in the 2000 edition.

16.   For an exhaustive recent survey of this doctrine see Ilaria L. E. Ramelli, *The
Christian Doctrine of Apokatastasis: A Critical Assessment from the New Testament to
Eriugena* (Leiden: Brill, 2013).

17.   For the Greek text and an English translation of *Ambiguum* 41 see Maximos
the Confessor, *On Difficulties in the Church Fathers: The "Ambigua,"* ed. N. Constas
(Cambridge, MA: Harvard University Press, 2014), 2:103–21. This *Ambiguum* is
also translated in Andrew Louth, *Maximus the Confessor* (London: Routledge, 1996),
156–62. Where given in the Constas edition, I have kept the references to the column numbers of the older edition in J. P. Migne, *Patrologia Graeca* 91.

18.   On Gregory of Nyssa's division of the sexes see E. Jeauneau, "La division

des sexes chez Grégoire de Nysse et chez Jean Scot Érigène," in *Eriugena: Studien zu seinen Quellen*, ed. W. Beierwaltes (Heidelberg: Carl Winter, 1980), 33–54; repr. in Jeauneau, *Études Érigéniennes*, 341–64.

19. See, e.g., *Periphyseon* 2.535A, CCCM 162:15; Sheldon-Williams transl., 135, with reference to Maximus, *Ambiguum* 37.

20. Although I am critical of Balthasar, my problem is not his suspected Hegelian inclinations, often adduced to keep his cosmological speculative views in check. See on this Joshua Lollar, "Reception of Maximian Thought in the Modern Era," in Allen and Neil, *Oxford Handbook of Maximus the Confessor*, 568–71. Whether or not one sees Balthasar as Hegelian, I see him as insufficiently Maximian.

21. This is powerfully stated in Peter Cramer, *Baptism and Change in the Early Middle Ages, c. 200–c. 1150* (Cambridge: Cambridge University Press, 1993), 7: "It is never just a spectacle, a priestly act observed by a detached congregation; but only happens at all because the congregation recognizes in the tension of symbol its own aspirations to be saved. It sees there the same mixture of the difficulty of attaining to truth, and the possibility of doing so." Cramer describes liturgy as "a search for meaning" (12).

22. See Louth, *Maximus the Confessor*, 155.

23. Louth, 155.

24. See Torstein T. Tollefsen, "Christocentric Cosmology," in Allen and Neil, *Oxford Handbook of Maximus the Confessor*, 314. The same point is made in Thomas Cattoi, "Liturgy as Cosmic Transformation," in Allen and Neil, *Oxford Handbook of Maximus the Confessor*, 419.

25. See Tollefsen, "Christocentric Cosmology," 315. I. P. Sheldon-Williams sees Maximus's triadic universe based more on the Porphyrian triad of being, power, and act (*ousia-dynamis-energeia*) than on the Plotinian-Proclean one of remaining-procession-conversion (*monè-proodos-epistrophè*) that Tollefsen sees Maximus adjusting. See I. P. Sheldon-Williams, "St. Maximus the Confessor," in *The Cambridge History of Later Greek and Early Medieval Philosophy*, ed. A. H. Armstrong (Cambridge: Cambridge University Press, 1967), 492–93.

26. See Tollefsen, "Christocentric Cosmology," 313.

27. See Tollefsen, 308, in reference to *Ambiguum* 33. For the text of that *Ambiguum* see Maximos the Confessor, *On Difficulties in the Church Fathers*, 2:63–65.

28. Thus, for example, Christ can be present both in nature and in scripture, as will be explained further below.

29. Maximos, *Ambiguum* 41, 105–7.

30. Maximos, 109–11.

31. Gregory of Nazianzus, *Oratio* 38.2, in Grégoire de Nazianze, *Discours 38–41*, ed. C. Moreschini and P. Gallay (Paris: Cerf, 1990), 106, lines 16–17.

32. Gregory of Nazianzus, *Oratio* 38.2, SC 358, 106, lines 9–10.

33. Maximos the Confessor, *Ambiguum* 31, in *On Difficulties*, 2:51.

34. Maximos the Confessor, 2:43.

35. I comment on Eriugena's use of this Maximian passage in "Creation and Epiphanic Incarnation: Reflections on the Future of Natural Theology from an Eriugenian-Emersonian Perspective," in *On Religion and Memory*, ed. B. S. Hellemans, W. Otten, and M. B. Pranger (New York: Fordham University Press, 2013), 76–79.

36. On Eriugena's eschatology toward the end of the *Periphyseon* see P. Dietrich and D. F. Duclow, "Virgins in Paradise: Deification and Exegesis in Periphyseon V," in *Jean Scot Écrivain: Actes du IVe Colloque International Montréal, 28 août–2 septembre, 1983*, ed. Guy-H. Allard (Montreal: Bellarmin / Paris: Vrin, 1986), 29–49.

37. See *Periphyseon* 3.723D, CCCM 163, 149, with reference to *Ambiguum* 10. See further M. Cappuyns, *Jean Scot Érigène: Sa vie, son œuvre, sa pensée* (Bruxelles: Culture et Civilisation, 1969), 276–80. See also D. F. Duclow, "Nature as Speech and Book in John Scotus Eriugena," *Mediaevalia* 3 (1977): 131–40.

38. Maximus starts his discussion at *Ambiguum* 10.17, in *On Difficulties* 1:191; Louth, *Maximus the Confessor*, 108.

39. Maximos, *Ambiguum* 10.18, in *On Difficulties* 1:197; Louth, *Maximus the Confessor*, 110.

40. The extended contemplation of the transfiguration commences at *Ambiguum* 10.31a, in *On Difficulties* 1:255; Louth, *Maximus the Confessor*, 128. The reading of Moses and Elijah with regard to time and nature can be found in *Ambiguum* 10.31a.9, in *On Difficulties* 1:263–65; Louth, *Maximus the Confessor*, 130–31.

41. See Dionysius, *Mystical Theology* 1.3, in *Pseudo-Dionysius: The Complete Works*, trans. Colm Luibheid (New York: Paulist Press, 1987), 136–37. Of course, Dionysius has a predecessor in Gregory of Nyssa's *Life of Moses*.

42. See Maximos, *Ambiguum* 10.31.e–g, in *On Difficulties* 1:271–73; Louth, *Maximus the Confessor*, 133. What we see here is a triadic manifestation of affirmative theology, negative theology being the ineffability of the God who is both monad and triad mentioned in 10.31.d. Triadic affirmative theology consists of activity, providence, and judgment. Activity is grounded in the beauty and magnificence of creation, indicating that God is the creator through the luminosity of the garments at the transfiguration, which signify visible creation.

43. For Maximus this is because in life Lazarus was without relevant ties to the world and without a healthy body.

44. See Maximos, *Ambiguum* 10.32, in *On Difficulties* 1:279–83; Louth, *Maximus the Confessor*, 135–37.

45. I leave out a further discussion of Maximus's *Mystagogy*. Since it builds on Dionysius's *Ecclesiastical Hierarchy*, it is more ecclesial than cosmic.

46. In this paragraph I draw on Paul M. Blowers, *Drama of the Divine Economy: Creator and Creation in Early Christian Theology and Piety* (Oxford: Oxford University Press, 2012), 159–66.

47. Blowers, 163.

48. Blowers, 163.

49. By *theosis* I refer to Athanasius's well-known statement in *On the Incarnation of the Word* 54.3 that "He [scil. Christ], indeed, assumed humanity that we might

become God." On Athanasius's Christology and *theosis* see Charles Kannengiesser, "Athanasius of Alexandria and the Foundation of Traditional Christology," *Theological Studies* 34, no. 1 (1973): 103–13; repr. in Charles Kannengiesser, *Arius and Athanasius: Two Alexandrian Theologians* (Hampshire: Variorum, 1991).

50. Other than on Hans Urs von Balthasar's *Cosmic Liturgy: The Universe according to Maximus the Confessor* and on the relevant texts in the Constas edition and translation, I have also drawn for this section on Andrew Louth, *Maximus the Confessor*, 63–77, including his partial translation of *Ambiguum 41* on pp. 155–63, as well as on the analysis of *Ambiguum 7* in Joshua Lollar, *To See into the Life of Things: The Contemplation of Nature in Maximus the Confessor and His Predecessors* (Turnhout: Brepols, 2013), 179–98.

51. Maximus will periodically refer back to this Gregorian opening passage. See also note 56 below.

52. Maximos, *On Difficulties* 1:82.

53. For Jean-Michel Garrigues, *Maxime le Confesseur: La charité, avenir divin de l'homme* (Paris: Beauchesne, 1976), *Ambiguum 7* offers Maximus's final word on divinization. In an attempt to avoid Origen's personalism and Gregory of Nyssa's epectasy, for Maximus one acquires a personal likeness in divinization that manifests itself as a passive return of the natural image into the archetype from which it came and in which it participates (97).

54. Maximos, *Ambiguum 7.20*, in *On Difficulties* 1:87. See also St. Maximus the Confessor, *On the Cosmic Mystery of Jesus Christ: Selected Writings from Maximus the Confessor*, trans. P. M. Blowers and R. L. Wilken (New York: St. Vladimir's Seminary Press, 2003), 50–51.

55. See Augustine, *Confessions* 1.1.

56. Maximos the Confessor, *Ambiguum 7.20*, in *On Difficulties* 1:100–102; St. Maximus the Confessor, *On the Cosmic Mystery of Jesus Christ*, 57–58. The expressions "portions of God" and "flowed down" derive from this *Ambiguum*'s opening quotation (see *On Difficulties* 1:74) from Gregory of Nazianzus's *Oratio* 14.7 (*PG* 35:865C). For the anti-Origenian thrust of this passage see Balthasar, *Cosmic Liturgy*, 133.

57. Maximos the Confessor, *Ambiguum 41.6*, in *On Difficulties* 2:108–10.

58. Maximos the Confessor, *Ambiguum 41.7*, in *On Difficulties* 2:110.

59. Virginia Burrus holds that under the influence of Nicea's strong Christology, the notion of humanity created in the image of God increasingly tilted toward the male, instituting it as normative. This baggage weighs as much on Maximus's view of the ascetic life as it does on the contemporary prohibition that women visit Mount Athos. See Virginia Burrus, *"Begotten, Not Made": Conceiving Manhood in Late Antiquity* (Stanford: Stanford University Press, 2000), 1–17.

60. In writing this chapter, I have been inspired by codirecting the dissertation of Bryce Rich, "Beyond Male and Female: Gender Essentialism and Orthodoxy" (PhD diss., University of Chicago, 2017).

CHAPTER 3

1. Of course the nature-grace opposition is a staple of scholarship on Augustine, with known highlights in the Pelagian debate (discussed below), as well as in the Jansenist controversy. For a different take on grace see Phillip Carey, *Inner Grace: Augustine in the Traditions of Plato and Paul* (Oxford: Oxford University Press, 2008).

2. See E. A. Clark, "'Adam's Only Companion': Augustine and the Early Christian Debate on Marriage," *Recherches augustiniennes* 21 (1986): 139–62. See also my essays "Augustine on Marriage, Monasticism and the Community of the Church," *Theological Studies* 59 (1998): 385–405; and "The Long Shadow of Human Sin: Augustine on Adam, Eve, and the Fall," in *Out of Paradise: Eve and Adam and Their Interpreters*, ed. B. E. J. H. Becking and S. A. Hennecke (Sheffield: Sheffield Phoenix Press, 2010), 29–49.

3. For the impact of Augustine on political thought see Arjo Vanderjagt, "Political Theology," in *The Oxford Guide to the Historical Reception of Augustine*, ed. K. Pollmann and W. Otten (Oxford: Oxford University Press, 2013), 3:1562–69. For the ramifications of Augustinian thought for modern politics and citizenship see Jean Bethke Elshtain, *Augustine and the Limits of Politics* (Notre Dame, IN: Notre Dame University Press, 1995); and Eric Gregory, *Politics and the Order of Love: An Augustinian Ethic of Democratic Citizenship* (Chicago: University of Chicago Press, 2008).

4. Augustine's important treatment of time in *Conf.* 11 is analyzed in P. Ricoeur, *Time and Narrative*, trans. Kathleen MacLaughlin and David Pellauer (Chicago: University of Chicago Press, 1984–88), 1:5–30 ("The Aporias of the Experience of Time: Book 11 of Augustine's Confessions"). For a critique of Ricoeur that applies Augustinian temporality to the reading of *Confessions*, see M. B. Pranger, *Eternity's Ennui: Temporality, Perseverance and Voice in Augustine and Western Literature* (Leiden: Brill, 2010), 38–54.

5. With this chapter's focus on the notion of "taking place" I intend both to invoke and criticize Jean-Luc Marion's *In the Self's Place: The Approach of Saint Augustine*, trans. Jeffrey L. Kosky (Stanford: Stanford University Press, 2012). While I applaud the way in which Marion's study pivots on the relation between God and self, my concern is that in doing so, he fails to make adequate room for creation. See my review of the French original in *Continental Philosophy Review* 42, no. 4 (2010): 597–602.

6. A cautionary note is sounded by Christopher Kirwan, who sees Augustine as defending free will to vindicate God as just punisher and caring creator and holds that, unlike Luther and Calvin, he maintained belief in the free decision of the will and never concluded to the bondage of the will. See Christopher Kirwan, *Augustine* (London: Routledge, 1991), 82. On sin and will see also James Wetzel, *Augustine: A Guide for the Perplexed* (New York: Continuum, 2010), 44–76 ("Sin and the Invention of Will"). On the complex role of corporeal existence and desire in Augustine see Virginia Burrus, Mark D. Jordan, and Karmen MacKendrick,

Seducing Augustine: Bodies, Desires, Confessions (New York: Fordham University Press, 2010).

7.   Found throughout *Conf.* 11, as in 11.23.30 ("I therefore see that time is some kind of extension") or 11.29.39 ("Look, my life is a stretch"). See note 4 above for Ricoeur's treatment of time and Pranger's critique.

8.   On Pelagius and Pelagianism see Peter Brown, *Augustine of Hippo: A Biography* (London: Faber and Faber, 1967), 340–52. Brown fittingly describes Pelagius as *emancipatus a Deo* (352), no longer dependent on the *pater familias*, with reference to Augustine, *Contra Iulianum opus imperfectum* 1:78.

9.   I see this as the consequence of the radicality of Pelagius's sense of emancipation, which I connect here with Peter Damian's later treatment of the question of whether God cannot just destroy what has been created but actually undo what has been done, for example, that Rome was never founded, in his *De divina omnipotentia* (1064 CE). I do so in part to point out that such debates occur in an ascetic context. On Damian's tract see Toivo J. Holopainen, *Dialectic and Theology in the Eleventh Century* (Leiden: Brill, 1996), 6–43.

10.   What I describe here as Pelagius's circularity may on a different level explain Augustine's difficulty in extracting himself from the Pelagian debate, which took up all his intellectual energy in the last decades of his life.

11.   See E. Gilson, "*Regio dissimilitudinis* de Platon à Saint Bernard de Clairvaux," *Mediaeval Studies* 9, no. 1 (1947): 108–30. The point I want to add is that in Augustine there is a certain convergence of place and time in concepts like *distentio animi* and *regio dissimilitudinis* that should not be overlooked. Rather than seeing these terms only in terms of sinfulness, I see them as his markers of a pressured ordinary life. Augustine's preference for ordinary life is the reason why, despite his focus on a historical paradise, he does not wax nostalgic about humanity's stay there prior to its ejection. If Augustine's rendition, say, of sex in paradise is stilted, it is in part because the exercise of a hypothetical reflection on utopian paradise is itself stilted.

12.   Robert A. Markus, *The End of Ancient Christianity* (Cambridge: Cambridge University Press, 1990), 45–62. Markus defines mediocrity in the context of Augustine's anti-Pelagian stance.

13.   Augustine, *Conf.* 10.33.50; Chadwick transl., 208. The phrase *mihi quaestio factus sum* occurs with some frequency in *Confessions*.

14.   Per the famous exchange in *Solil.* 1.2.7: *scire cupio Deum et animam*. See Saint Augustine, *Soliloquies and Immortality of the Soul*, with an introduction, translation, and commentary by G. Watson (Warminster UK: Aris and Phillips, 1990), 30–31.

15.   Mark Vessey, ed., *A Companion to Augustine* (Oxford: Wiley-Blackwell, 2012), lacks a chapter on creation, while Scott Dunham, *The Trinity and Creation in Augustine: An Ecological Analysis* (Albany: State University of New York Press, 2008), subsumes creation under God. I have discussed Dunham in W. Otten, "Nature as a Theological Problem: An Emersonian Response to Lynn White," in *Responsibility*

and the Enhancement of Life: Essays in Honor of William Schweiker, ed. G. Thomas and H. Springhart (Leipzig: Evangelische Verlagsanstalt, 2017), 265–80.

16. Simo Knuuttila, "Time and Creation in Augustine," in *The Cambridge Companion to Augustine*, ed. E. Stumpf and N. Kretzmann (Cambridge: Cambridge University Press, 2001), 103–15.

17. Paul M. Blowers, *Drama of the Divine Economy: Creator and Creation in Early Christian Theology and Piety* (Oxford: Oxford University Press, 2012), 153–59.

18. Augustine, *Conf.* 1.1.1; Chadwick transl., 3:

> "You are great, Lord and highly to be praised (Ps. 47:2): great is your power and your wisdom is immeasurable" (Ps. 146:5). Man, a little piece of your creation (*aliqua portio creaturae tuae*), desires to praise you, a human being, "bearing his mortality with him" (2 Cor. 4:10), carrying with him the witness of his sin and the witness that you "resist the proud" (1 Pet. 5:5). Nevertheless to praise you is the desire of man, a little piece of your creation (*aliqua portio creaturae tuae*). You stir man to take pleasure in praising you, because you have made us for yourself (*quia fecisti nos ad te*), and our heart is restless until it rests in you. "Grant me Lord to know and understand" (Ps. 118:34, 73, 144) which comes first—to call upon you or to praise you and whether knowing you precedes calling upon you.

19. Augustinus, *Retract.* 2.6.1, ed. Almit Mutzenbecher, CCSL 57 (Turnhout: Brepols, 1984): "a primo usque ad decimum de me scripti sunt, in tribus ceteris de scripturis sanctis, ab eo quod scriptum est: 'in principio fecit deus caelum et terram,' usque ad sabbati requiem."

20. Sabine MacCormack, "Augustine Reads Genesis," *Augustinian Studies* 39, no. 1 (2008): 5–47.

21. See Augustinus, *The Literal Meaning of Genesis* 1.15.29, in Saint Augustine, *On Genesis*, trans. Edmund Hill, O.P. (New York: New City Press, 2002), 181. I have consulted the following Latin text: *La Genèse au sens littéral en douze livres: De Genesi ad litteram libri duodecim*, traduction, introduction et notes par P. Agaësse et A. Solignac, vol. 1 (bks. 1–7); vol. 2 (bks. 8–12), Œuvres de St. Augustin 48–49 (Paris: Desclée de Brouwer, 1972).

22. MacCormack, "Augustine Reads Genesis," 35.

23. For a fuller treatment of Augustinian exegesis prior to *De Genesi ad litteram*, see Michael Cameron, *Christ Meets Me Everywhere: Augustine's Early Figurative Exegesis* (Oxford: Oxford University Press, 2012). On the diverse aspects of Augustinian exegesis see P. Bright, ed. and trans., *Augustine and the Bible* (Notre Dame, IN: University of Notre Dame Press, 1999).

24. Augustine, *The Literal Meaning of Genesis*, 181. For the Latin text of the latter half of this quotation, see Augustine, *Gen. litt.* 1.15.29; Agaësse-Solignac edn.: 1:120: *sed quia illud, unde fit aliquid, etsi non tempore, tamen quadam origine prius est, quam illud quod inde fit, potuit diuidere scriptura loquendi temporibus quod deus faciendi temporibus non diuisit.* MacCormack (34–35) translates the passage slightly differently: "But that from which something is made, even if it is not prior

in time to that which is made from it, is yet prior in some manner of origin; therefore Scripture in the time of its speaking could separate what God in the time of his making did not separate."

25.   On Origen see Chap. 2n2 herein.

26.   See Brian E. Daley, "Origen's *De principiis*: A Guide to the Principles of Christian Scriptural Interpretation," in *Nova et vetera: Patristic Studies in Honor of Thomas Patrick Halton*, ed. J. Petruccione (Washington: Catholic University of America Press, 1998), 6.

27.   For Augustine's exploration of the literal sense in *Gen. litt.* see Charles Kannengiesser, *Handbook of Patristic Exegesis* (Leiden: Brill, 2004), 2:1165–69.

28.   Eusthatius Afer is said to have translated Basil's homilies into Latin around 400 CE. They are supposed to have been translated again in the sixth century by Dionysius Exiguus.

29.   See Philip Rousseau, "Human Nature and Its Material Setting in Basil of Caesarea's Sermons on the Creation," *Heythrop Journal* 49 (2008): 222–39.

30.   See Rousseau, 223n12. See also Richard Lim, "The Politics of Interpretation in Basil of Caesarea's Hexaemeron," *Vigiliae Christianae* 44, no. 4 (1990): 315–70.

31.   Rousseau, "Human Nature," 223.

32.   In Basil, at least in Rousseau's reading of him, the question of whether speaking or writing is attributed to Scripture (or Moses) does not make a difference.

33.   This is true even if, as Rousseau adds, that reader is inspired by the Spirit and alert to the presence of Christ; see Rousseau, "Human Nature," 225.

34.   Rousseau, "Human Nature," 226. Here one may think, for example, of Gregory of Nyssa's notion of epectasy in reference to Phil 3:13, where we are urged to "forget what lies behind and strain forward to what lies ahead." See on this J. Warren Smith, *Passion and Paradise: Human and Divine Emotion in the Thought of Gregory of Nyssa* (New York: Herder-Crossroad, 2004), 104–25.

35.   Rousseau, "Human Nature," 232.

36.   Rousseau, 231.

37.   For the broader anthropological context of Eastern Christianity with attention to Origen and Maximus as well, see Lars Thunberg, "The Human Person as Image of God: I. Eastern Christianity," in *Christian Spirituality: Origins to the Twelfth Century*, ed. Bernard McGinn and John Meyendorff (New York: Crossroad, 1988), 291–312.

38.   For the broader anthropological context of Western Christianity in which Augustine is embedded, see Bernard McGinn, "The Human Person as Image of God: I. Western Christianity," In *Christian Spirituality: Origins to the Twelfth Century*, ed. Bernard McGinn and John Meyendorff (New York: Crossroad, 1988), 312–30.

39.   Karla Pollmann, "Augustine, Genesis, and Controversy," *Augustinian Studies* 38, no. 1 (2007): 203–16.

40.   Pollmann, 210. In Gn 1:3–4 light is neither material nor metaphorical but spiritual.

41. Pollmann, 210.

42. This is a point not granted by Pollmann, who states, "Generally, Augustine, like other ecclesiastical writers, is of the opinion that biblical exegesis is not really able to tell anything fundamentally new; it rather serves to educate the readers of the Bible in a moral and/or intellectual way" (212). In this she accords with Knuuttila and Blowers (see above nn. 16 and 17) rather than MacCormack in not granting a fundamental creativity to Augustine's reading of scripture. Precisely that creativity is what I want to retrieve as a crucial factor in his decidedly exegetical view of nature and creation. Given that God creates everything at once (*omnia simul*), scripture is needed to paint the details of creation with words.

43. See Jean-Luc Marion, "Resting, Moving, Loving: The Access to the Self according to Saint Augustine," in "The Augustinian Moment," ed. Willemien Otten, special issue, *Journal of Religion* 91, no. 1 (2011): 28.

44. See Marie-Anne Vannier, "Le rôle de l'hexaéméron dans l'interprétation augustinienne de la création," *Revue des sciences philosophiques et théologiques* 71 (1987): 538, with reference to Gerhart B. Ladner, *The Idea of Reform: Its Impact on Christian Thought and Action in the Age of the Fathers*, rev. ed. (New York: Harper and Row, 1967), 167. The English translation of the French is my own. Vannier elaborated on her thesis in her subsequent monograph, *"Creatio," "Conversio," "Formatio" chez S. Augustin* (Paris: Beauchesne, 1991), 73–82, for the development of this scheme.

45. In *Augustine's Inner Dialogue: The Philosophical Soliloquy in Late Antiquity* (Cambridge: Cambridge University Press, 2010) Brian Stock draws a line from the ancient tradition of inner dialogue to the innovative Augustinian soliloquy, with the latter allowing for the thematization of self-existence and personal and civilizational history (230). I would add the history of creation as a possible third narratival context.

46. Augustine, *Literal Meaning of Genesis*, 171.

47. Augustine, 186.

48. Augustine, *Conf.* 13.29.44; Chadwick transl., 300.

49. Andrea Nightingale, *Once Out of Nature: Augustine on Time and the Body* (Chicago: University of Chicago Press, 2011), 23–54 ("Eden and Resurrected Transhumans").

50. See the critical review by Burcht M. Pranger in *Journal of Religion* 92, no. 3 (2012): 423–24, which points to the category mistake of dividing time by seeing it as both psychic, as does Augustine, and earthly, which is Nightingale's own term, resulting in a muddled sense of temporality. James Wetzel states that earthly time is not a form of time; see James Wetzel, review of *Once Out of Nature*, by Andrea Nightingale, *Notre Dame Philosophical Reviews: An Electronic Journal*, Jan. 4, 2012, http://ndpr.nd.edu/news/once-out-of-nature-augustine-on-time-and-the-body.

51. In *Periphyseon* 4.808D–9A Eriugena concludes from Augustine's comment in *De ciuitate Dei* 14.26.1–3 "that man began to live (*uiuebat*) rather than lived or had lived (*uixit* or *uixerat*) in paradise in the enjoyment of God without any want"

that the use of the inceptive imperfect signals the beginning of some action that had by no means reached completion. Compare to this my comments on Adam's brief stay in paradise in Bernard Silvestris's *Cosmographia* in Otten, *From Paradise to Paradigm: A Study of Twelfth-Century Humanism* (Leiden: Brill, 2004), 21–23.

52. See the reviews by Pranger and Wetzel on this point (note 50 above).

53. See Otten, "Augustine on Marriage," 398, with reference to *De bono coniugali* 1.1. See, furthermore, *De ciuitate Dei* 14.11, where Augustine sees Adam's sin as partly social in that he did not stand up to Eve's suggestion: "It was the same with that first man and his wife. They were alone together, two human beings, a married pair; and we cannot believe that the man was led astray to transgress God's law because he believed that the woman spoke the truth, but that he fell in with her suggestions because they were so closely bound in partnership." Augustine, *Concerning the City of God against the Pagans*, trans. H. Bettenson (London: Penguin, 1972), 570.

54. See on this my essays "Augustine on Marriage," and "The Long Shadow of Human Sin."

55. Augustine, *Conf.* 10.27.38; Chadwick transl., 201: "Late have I loved you, beauty so old and so new: late have I loved you. And see, you were within and I was in the external world and sought you there, and in my unlovely state I plunged into those lovely created things which you made."

56. Augustine, *Conf.* 8.7.17; Chadwick transl., 145: "But I was an unhappy young man, wretched as at the beginning of my adolescence when I prayed you for chastity and said: 'Grant me chastity and continence, but not yet.' I was afraid you might hear my prayer quickly, and that you might too rapidly heal me of the disease of lust which I preferred to satisfy rather than suppress."

57. William Butler Yeats, "Sailing to Byzantium" (1928), in *The Collected Poems of W. B. Yeats*, ed. Richard Finneran (New York: Simon and Schuster, 1989).

58. Vannier, "Le rôle de l'hexaéméron," 543: "Pour exprimer la création, Augustin met l'accent sur l'anthropologie, sur la relation au Créateur."

59. Augustine, *Literal Meaning of Genesis*, Hill transl., 464–506.

60. Augustine, *Gen. litt.* 12.37.70; Hill transl., 506.

61. Augustine, *Gen. litt.* 12.34.67; Hill transl., 504.

62. Augustine, *Gen. litt.* 1.2.5; Hill transl., 170.

POSTSCRIPT TO PART I

1. Rom 1:20 (RSV): "Ever since the creation of the world his invisible nature, namely, his eternal power and deity, has been clearly perceived in the things that have been made." Next to the transfiguration scene, this is another biblical trope that underlies the parallelism of nature and scripture treated in Chapter 2. See also W. Otten, "Nature and Scripture: Demise of a Medieval Analogy," *Harvard Theological Review* 88, no. 2 (1995): 257–84.

2. Augustine, *Conf.* 9.10.25, ed. O'Donnell, 1:113–14; Chadwick transl., 171–72.

3. Augustine, *Conf.* 10.6.9, ed. O'Donnell, 1:122; Chadwick transl., 183. Cf. Robert A. Markus, *Signs and Meanings: World and Text in Ancient Christianity* (Eugene, OR: Wipf and Stock, 2011), "World and Text I," 1–43, 27.

4. See Markus, *Signs and Meanings*." While I have drawn mostly on Markus's first chapter, the book contains three other articles on signs and semiotic theory in Augustine as well. For an overview of Augustine's thought on signification and language see also John M. Rist, *Augustine: Ancient Thought Baptized* (Cambridge: Cambridge University Press, 1994), 23–40.

5. Markus, *Signs and Meanings*, "World and Text I," 19, with reference to O'Donnell, 3:316.

6. Markus, 11.

7. Markus, 10.

8. In *De ciuitate Dei* 11.18 Augustine compares the verbal beauty of antithesis or opposition in a poem to the beauty in "the composition of the world's history arising from the antithesis of contraries—a kind of eloquence in events, instead of in words" (Bettenson transl., 449). Following Markus, I broaden the notion of *res* (things) in *eloquentia rerum* to include natural things, as well as historical, prophetic events.

9. The evangelist Luke is traditionally depicted as a bull or calf, whereas Mark is a lion, Matthew a man, and John an eagle. See Thomas P. Scheck, trans., *St. Jerome: Commentary on Matthew* (Washington, DC: Catholic University of America Press, 2008). These identifications fill out various biblical texts, esp. Rv 4:2, 6–8 and Ez 1:5–10.

10. Markus, *Signs and Meanings*, 11.

11. Markus, 29.

12. See Markus, 30.

13. See Elizabeth Kendig, "La forme dialogique dans le *Periphyseon*: Recréer l'esprit," *Les études philosophiques*, no. 1 (Jan. 2013): 101–19.

14. Eriugena states in *Periphyseon* 5.893C, CCCM 165:49; Sheldon-Williams transl., 562: "In man every creature is established (*condita*), both visible and invisible. Therefore he is called the workshop of all (*officina omnium*), seeing that in him all things that came after God are contained. Hence he is also customarily called the Intermediary (*medietas*), for since he consists in soul and body he comprehends within himself and gathers into one two ultimate extremes of the spiritual and the corporeal." Note that *condita* can also mean created. Maximus discusses the notion of humanity as workshop in the context of his five divisions, discussed in Chap. 2; see *Ambiguum* 41, in *On Difficulties* 2:104.

15. *Periphyseon* 4.768B, CCCM 164:40; Sheldon-Williams transl., 413: "We may then define man as follows: Man is a certain intellectual concept formed eternally in the Mind of God."

16. *Periphyseon* 4.768C–D, CCCM 164:40–41; Sheldon-Williams transl., 414: "and indeed the essence of man is considered principally to consist in this: that it has been given him to possess the concept of all things which were either created his equals or which he was instructed to govern. For how could man be given the dominion of things of which he had not the concept?"

17. See W. Otten, *Anthropology of Johannes Scottus Eriugena* (Leiden: Brill, 1991), 203–19.

18. Auerbach's famous essay "Odysseus' Scar" characterizes biblical texts, as opposed to classical ones, as "fraught with background." This places a need for interpretation on such texts and a duty to tease out their moral truth on their readers as they translate them into contemporaneous cultural categories. See Erich Auerbach, *Mimesis: The Representation of Reality in Western Literature*, trans. Willard R. Trask (Princeton, NJ: Princeton University Press, 1968), 3–23.

19. *Periphyseon* 4.841D–42A, CCCM 164:142–43; Sheldon-Williams transl., 500–501.

20. See W. Otten, "The Long Shadow of Human Sin: Augustine on Adam, Eve, and the Fall," in *Out of Paradise: Eve and Adam and Their Interpreters*, ed. B. E. J. H. Becking and S. A. Hennecke (Sheffield: Sheffield Phoenix Press, 2010), 36, with reference to *De ciuitate Dei* 14.11.

21. *Periphyseon* 4.846B–C, CCCM 164:148–49; Sheldon-Williams transl., 506.

22. See W. Otten, "The Dialectic of the Return in Eriugena's *Periphyseon*," *Harvard Theological Review* 84 (1991): 409.

23. *Periphyseon* 5.862A–C, CCCM 165:5; Sheldon-Williams transl., 525.

24. Stanley Cavell, *Conditions Handsome and Unhandsome: The Constitution of Emersonian Perfectionism* (Chicago: University of Chicago Press, 1990), xxx.

25. *Periphyseon* 5.862B–C, CCCM 165:5; Sheldon-Williams transl., 525.

CHAPTER 4

1. See Richard Crouter's introduction to Friedrich Schleiermacher, *On Religion: Speeches to Its Cultured Despisers*, trans. and ed. Richard Crouter (Cambridge: Cambridge University Press, 2015), xvii.

2. See Crouter's introduction, xi: "At the time of his death Schleiermacher was the most distinguished theologian of Protestant Germany, the author of a modern, post-Enlightenment system of theology that ranks with Calvin and Aquinas in the history of Christian thought." To this list Augustine is added in Jacqueline Mariña, ed., *The Cambridge Companion to Friedrich Schleiermacher* (Cambridge: Cambridge University Press, 2005), 1. For the *Speeches* see *Über die Religion: Reden an die Gebildeten unter ihren Verächtern (1799)*, ed. Günter Meckenstock (Berlin: De Gruyter, 1984), which I cite according to the text of the *Kritische Gesamtausgabe* (*KGA*) 1.2, *Schriften aus der Berliner Zeit, 1796–1799*, 189–326; and Crouter's translation.

3. Emerson appears to have been exposed to Schleiermacherian influence, and

comparisons with the "Address" have been especially noted. See Robert D. Richardson Jr., "Schleiermacher and the Transcendentalists," in *Transient and Permanent: The Transcendentalist Movement and Its Contexts*, ed. Charles Capper and Conrad E. Wright (Boston: Massachusetts Historical Society, 1999), 121–47. In other ways, however, Schleiermacher is more akin to William James. I address their affinity in Chapter 5.

4. Published in 1793, Kant's *Religion within the Limits of Reason Alone* is marked by the repressive culture in which Prussia forbade him from publishing further works on religion. After the death of the king he published an expanded larger edition. See Crouter's introduction, xv. The role and pressures of censorship are not to be underestimated. See on Schleiermacher's decision to publish the *Speeches* anonymously and his exchanges with his church superior Fr. S. G. Sack, who censored them, Julia S. Lamm, *The Living God: Schleiermacher's Theological Appropriation of Spinoza* (University Park: Pennsylvania State University Press, 1996), 58.

5. See Marcia L. Colish, *The Mirror of Language: A Study in the Medieval Theory of Knowledge* (Lincoln: University of Nebraska Press, 1983), 16: "Augustine projected a redeemed rhetoric as the outcome of a revealed wisdom." See also W. Otten, "The Tension between Word and Image in Christianity," in *Iconoclasm and Iconoclash: Struggle for Religious Identity*, ed. P. van Geest, D. Müller, W. van Asselt, and Th. Salemink (Leiden: Brill, 2007), 33–36.

6. I make this argument in "Creation and Epiphanic Incarnation: Reflections on the Future of Natural Theology from an Eriugenian-Emersonian Perspective," in *On Religion and Memory*, ed. B. S. Hellemans, W. Otten, and M. B. Pranger (New York: Fordham University Press, 2013), 64–74.

7. See Maïeul Cappuyns, *Jean Scot Érigène: Sa vie, son œuvre, sa pensée* (Bruxelles: Culture et Civilisation, 1969).

8. Lamm, *The Living God*, 1, per the reaction from the censor Fr. S. G. Sack. Schleiermacher's letter to Jacobi is dated March 30, 1818. See also Chap. 1nn23–25 herein.

9. See Lamm, *The Living God*, 21.

10. See Lamm, 23–24. Translated excerpts from Jacobi's text, *Über die Lehre des Spinoza, in Briefen an den Herrn Moses Mendelssohn* can be found in Gérard Vallée, *The Spinoza Conversations between Lessing and Jacobi: Texts with Excerpts from the Ensuing Controversy*, introd. Gérard Vallée, trans. G. Vallée, J. B. Lawson, and C. G. Chapple (Lanham, MD: University Press of America, 1988), 78–125.

11. Jacobi, *Über die Lehre des Spinoza*, 87.

12. Jacobi, 87.

13. Jacobi, 88.

14. Lamm, *The Living God*, 25.

15. See Lamm, 26.

16. Wolfhart Pannenberg, *The Historicity of Nature: Essays on Science and*

Theology, ed. Niels Henrik Gregersen (West Conshohocken, PA: Templeton Foundation Press, 2008), 18.

17. At this point I leave Lamm's analysis behind but not without mentioning that she discusses Schleiermacher's religion as that which gives system. See Lamm, *The Living God*, 60.

18. See Richardson, "Schleiermacher and the Transcendentalists."

19. In "The Transcendentalist," *CW* 1:213, Emerson speaks of a double consciousness, of understanding and the soul, as our two lives that "show very little relation to each other." One is "all buzz and din," the other "all infinitude and paradise." I have connected this to Eriugena's notion of a *duplex theoria* in W. Otten, "Nature as a Theological Problem: An Emersonian Response to Lynn White," in *Responsibility and the Enhancement of Life: Essays in Honor of William Schweiker*, ed. Günter Thomas and Heike Springhart (Leipzig: Evangelische Verlagsanstalt, 2017), 278–79. On Eriugena's *duplex theoria* see W. Beierwaltes, "Duplex Theoria: Zu einer Denkform Eriugenas," in *Begriff und Metapher: Sprachform des Denkens bei Eriugena*, ed. W. Beierwaltes (Heidelberg: Carl Winter, 1990), 39–64.

20. I generally follow the first edition here. For a note on the editions see Crouter, introduction, xliv–xlv.

21. On Schleiermacher's echoing of the Romantic predilection for geometric patterns see Martin Jay, *Songs of Experience: Modern American and European Variations on a Universal Theme* (Berkeley: University of California Press, 2005), 94.

22. Friedrich Schleiermacher, "First Speech: Apology," in *On Religion*, 3–4 (my italics); *KGA* 1.2, 189–90 (my italics).

23. See Justin Martyr, "The First Apology of Justin, the Martyr," 1.9–10, in *Early Christian Fathers*, ed. Cyril C. Richardson (New York: Touchstone, 1996), 246–47.

24. Schleiermacher, "First Speech: Apology," 5; *KGA* 1.2, 191: "es ist die innere unwiderstehliche Notwendigkeit meiner Natur, es ist ein göttlicher Beruf, es ist das was meine Stelle im Universum bestimmt, und mich zu dem Wesen macht, welches ich bin."

25. Schleiermacher, 5; *KGA* 1.2, 191.

26. Schleiermacher, 6; *KGA* 1.2, 192: "ein allgemeines Band des Bewusstseins."

27. Schleiermacher, 6; *KGA* 1.2, 192.

28. Schleiermacher, 6–7; *KGA* 1.2, 193–94.

29. Schleiermacher, 8; *KGA* 1.2, 194. While Schleiermacher sounds in some ways Emersonian here, there is a sequential coloring to the role of mediator that is absent in Emerson.

30. Schleiermacher, 10; *KGA* 1.2, 196: "wo dieser heilige Instinkt verborgen liegt."

31. Schleiermacher, 10; *KGA* 1.2, 197: "in die innerste Tiefen möchte ich Euch geleiten, aus denen sie zuerst das Gemüth anspricht."

32. Friedrich Schleiermacher, "Second Speech: On the Essence of Religion," in

On Religion: Speeches to Its Cultured Despisers, trans. and ed. Richard Crouter (Cambridge: Cambridge University Press, 2015), 33; *KGA* 1.2, 223.

33. Crouter usefully lays out the rhetorical scheme of the *Speeches* in his introduction (Schleiermacher, *On Religion*, xxxi), even as he also clarifies that Schleiermacher's *Speeches* are not about what to believe but rather about what it means to believe. While the *Speeches* are, indeed, not doctrinal, they are about religion and not, say, about faith or religious experience, which will be James's topic.

34. Friedrich Schleiermacher, "Fifth Speech: On the Religions," in *On Religion: Speeches to Its Cultured Despisers*, trans. and ed. Richard Crouter (Cambridge: Cambridge University Press, 2015), 110; *KGA* 1.2, 310–11. I have been influenced by the argument of an unpublished paper by Peter de Mey (University of Louvain), "The Defence of Revealed Christianity through an Appeal to Experience in Hume, Lessing, and Schleiermacher," on the point that even before Schleiermacher, though he was likely unaware, Lessing and Hume also held to the superiority of revealed religion over natural religion. De Mey argues that Schleiermacher is dissatisfied with contemporary theology and with deistic solutions but warns to take the spirit of religion seriously. The text of De Mey's presentation for the 2001 AAR Schleiermacher group can be found here: www.academia.edu/14382595/The_Defence _of_Revealed_Christianity_through_an_Appeal_to_Experience_in_Hume_Lessing _and_Schleiermacher.

35. As Jacqueline Mariña remarks about *The Christian Faith*, indicating a next step in this development: "Many of Schleiermacher's critics have concluded that *The Christian Faith* presents an anthropological transcendental philosophy of religion with an amazingly high Christology stuck in the middle." See her "Christology and Anthropology in Friedrich Schleiermacher," in Mariña, *Cambridge Companion*, 159. See contrary to this, however, Andrew C. Dole, *Schleiermacher on Religion and the Natural Order* (Oxford: Oxford University Press, 2010). Dole argues that "Schleiermacher's complete account of the contents of Christianity amounts to a 'supplemental naturalism': an understanding of Christianity as a natural phenomenon combined with claims that go beyond what the sciences have or can establish" (153).

36. I borrow the term *excarnation* from Charles Taylor, *A Secular Age* (Cambridge, MA: The Belknap Press of Harvard University Press, 2007), 613–14, but I do not adopt his ecclesial reading. Taylor defines *excarnation* as the transfer of our religious life out of bodily forms of ritual, worship, practice, so that it more and more comes to reside "in the head." My use is more in line with what Taylor sees as another facet of excarnation—namely, "what disengaged reason demands of desire." Excarnation refers to a thinning out of nature, an unfleshing in favor of a logical reduction to causality, yet without the relationality that characterizes the Jamesian cosmos.

37. Hans Joas, *Die Macht des Heiligen: Eine Alternative zur Geschichte von der Entzauberung* (Berlin: Suhrkamp, 2017), 24–60, 61–84.

38. But see note 34 above for a caveat on Hume, which is echoed by Joas, *Die Macht des Heiligen*, 58–59.

39. Joas, 56.

40. Joas, 59.

41. See Dole, *Schleiermacher on Religion*, 155.

42. In all fairness, for Schleiermacher causality is obviously not a contraction but represents that the fullness of the divine attributes must be rooted in divine causality, since they only explain the feeling of divine dependence. See Robert M. Adams, "Schleiermacher on Evil," *Faith and Philosophy* 13, no. 4 (1996): 563–64.

43. Schleiermacher, "Second Speech," 46; *KGA* 1.2, 237: "Alle diese Gefühle sind Religion, und ebenso alle andere, bei denen das Universum der eine, und auf irgend eine Art Euer eignes Ich der andere von den Punkten ist zwischen denen das Gemüth schwebt."

44. Schleiermacher, 52; *KGA* 1.2, 244.

45. See, e.g., Schleiermacher, "First Speech," 6; *KGA* 1.2, 192 ("jenen geschlossenen Ring").

46. Robert M. Adams, "Faith and Religious Knowledge," in *The Cambridge Companion to Friedrich Schleiermacher*, ed. Jacqueline Mariña (Cambridge: Cambridge University Press, 2005), 35–51.

47. See Wayne Proudfoot, *Religious Experience* (Berkeley: University of California Press, 1985), 7, 11, 237n7, cited in Adams, "Faith and Religious Knowledge," 36.

48. Schleiermacher, "Second Speech," 24; *KGA* 1.2, 213. Cited in Adams, "Faith and Religious Knowledge," 36.

49. Schleiermacher, "Second Speech," 25; *KGA* 1.2, 214. Cited in Adams, "Faith and Religious Knowledge," 36.

50. Schleiermacher, 25; *KGA* 1.2, 214.

51. Cited in Adams, "Faith and Religious Knowledge," 37.

52. *Der christliche Glaube* §4.3, as cited in Adams, "Faith and Religious Knowledge," 37.

53. Schleiermacher, "Second Speech," 19; *KGA* 1.2, 207: "Stellet Euch auf den höchsten Standpunkt der Metaphysik und der Moral, so werdet Ihr finden, dass beide mit der Religion denselben Gegenstand haben, nämlich das Universum und das Verhältnis des Menschen zu ihm." Adams refers to this as an instantiation of a preconceptual object (see note 48 above), but can one not read religion here as having a clearly defined object in the universe and the human relation to it, but one that is not unique to religion? Clearly, Schleiermacher continues the second speech by separating religion from metaphysics and morality, and sees contamination between them as a problem, but that does not detract from the antecedent validity of this observation.

CHAPTER 5

1. See Robert D. Richardson, "Schleiermacher and the Transcendentalists," in *Transient and Permanent: The Transcendentalist Movement and Its Contexts*, ed. Charles Capper and Conrad E. Wright (Boston: Massachusetts Historical Society, 1999), 121–47.

2. See Charles Taylor, *Varieties of Religion Today: William James Revisited* (Cambridge, MA: Harvard University Press, 2002).

3. Martin Jay, *Songs of Experience: Modern American and European Variations on a Universal Theme* (Berkeley: University of California Press, 2005), 93, 100–101. It appears Schleiermacher can hardly be introduced without mention being made of Barth's criticism.

4. See Charles Taylor, *A Secular Age* (Cambridge, MA: Belknap Press of Harvard University Press, 2007). Taylor's embrace of Catholicism comes out at various points, for example, in the chapter on conversions (728–72).

5. See William James, *The Varieties of Religious Experience* (Cambridge, MA: Harvard University Press, 1985). I quote by page number in the main text.

6. Peter Harrison points to the new status of religion in modernity, where it calls on science to adjudicate instead of being asked by scientists to adjudicate. This leads to a new role for apology in Locke and, I would argue, also in Schleiermacher. See Peter Harrison, "Miracles, Early Modern Science, and Rational Religion," *Church History* 75, no. 3 (2006): 493–510.

7. See Jay, *Songs of Experience*, 109–10, on the critical reception of James. As an example, Daniel L. Pals, *Seven Theories of Religion* (New York: Oxford University Press, 1996), a standard survey text, makes no mention of James but does have chapters on Durkheim, Marx, and Geertz (233–67). See also Robert N. Bellah, *Religion in Human Evolution: From the Paleolithic to the Axial Age* (Cambridge, MA: Belknap Press of Harvard University Press, 2011). On Geertz see pp. xiv–xvii.

8. For very different examples of making religion part of scientific analysis, see David S. Wilson, *Darwin's Cathedral: Evolution, Religion, and the Nature of Society* (Chicago: University of Chicago Press, 2003), who follows an evolutionary approach; Ann Taves, *Fits, Trances, and Visions: Experiencing Religion and Explaining Experience from Wesley to James* (Princeton, NJ: Princeton University Press, 1999); and Ann Taves, *Religious Experience Reconsidered: A Building-Block Approach to the Study of Religion and Other Special Things* (Princeton, NJ: Princeton University Press, 2009), who is more interested in neurological information.

9. See David C. Lamberth, *William James and the Metaphysics of Experience* (Cambridge: Cambridge University Press, 1999), 4, on relations as part of religious experience.

10. William James, "The Will to Believe," in *The Will to Believe and Other Essays in Popular Philosophy* (Cambridge, MA: Harvard University Press, 1979), 13–33, cited by page number in the main text.

11. See Michael R. Slater, *William James on Ethics and Faith* (Cambridge: Cambridge University Press, 2009), 1–16.

12. The reference is to William Kingdon Clifford (1845–79), British mathematician and philosopher, who held that it is wrong to accept a belief when the evidence is insufficient even in case the belief is true. See James, *The Will to Believe*, 255–56.

13. William James, "The Sentiment of Rationality," in *The Will to Believe*, 57–89, which I quote likewise by page number.

14. See on this Leszek Kolakowski, *God Owes Us Nothing: A Brief Remark on Pascal's Religion and on the Spirit of Jansenism* (Chicago: University of Chicago Press, 1995), 3–110.

15. James realizes his exaggeration when he says: "Surely Pascal's own personal belief in masses and holy water had far other springs; and this celebrated page of his is but an argument for others, a last desperate snatch at a weapon against the hardness of the unbelieving heart" ("Will to Believe," 16).

16. Aside from my comments on Pascal's wager in James, Michael Slater sees James himself deploying "a Pascalian prudential argument for religious belief" that he labels James's wager but sees ultimately fail. See Slater, *William James on Ethics*, 48–66.

17. Lamberth, *William James*, 31.

18. In 1904–5 James published "Is Radical Empiricism Solipsistic?," rejecting the criticism that it is. See William James, *Essays in Radical Empiricism* (Cambridge, MA: Harvard University Press, 1976), 119–22.

19. Wayne Proudfoot, "William James on an Unseen Order," *Harvard Theological Review* 93, no. 1 (2000): 66.

20. Proudfoot, 65.

21. James, *Varieties*, 51. Quoted in Wayne Proudfoot, "Pragmatism and 'An Unseen Order' in *Varieties*," in *William James and a Science of Religions: Reexperiencing "The Varieties of Religious Experience,"* ed. Wayne Proudfoot (New York: Columbia University Press, 2004), 32.

22. Proudfoot, "Pragmatism," 43.

23. See James, *Varieties*, 359, cited in Proudfoot, 43.

24. See Richard Rorty, "Some Inconsistencies in James's *Varieties*," in *William James and a Science of Religions: Reexperiencing "The Varieties of Religious Experience,"* ed. Wayne Proudfoot (New York: Columbia University Press, 2004), 91, 95–96.

25. Proudfoot, "Pragmatism," 44.

26. The substance of the essay was first published in *Mind* in 1879 and then in the *Princeton Review*, before being reissued in *The Will to Believe and Other Essays in Popular Philosophy* in 1897. I follow the latter version here, as reprinted in the 1979 Harvard University Press edition (57–89), which will be cited by page number in the main text. See also James, *The Will to Believe*, 326–33.

27. The critical edition (264) references Emerson, *Letters and Social Aims* (Boston: Houghton, Mifflin, 1883), 287; vol. 8 of the Riverside Edition of *Emerson's*

Complete Works was in James's possession. On James's pragmatic reading of Emerson see James M. Albrecht, *Reconstructing Individualism: A Pragmatic Tradition from Emerson to Ellison* (New York: Fordham University Press, 2012), 25–52.

28. This would make for a very Augustinian quest, per my analysis in Chap. 3 above.

29. Ann Taves, "The Fragmentation of Consciousness and *The Variety of Religious Experience*: William James's Contribution to a Theory of Religion," in *William James and a Science of Religions: Reexperiencing "The Varieties of Religious Experience,"* ed. Wayne Proudfoot (New York: Columbia University Press, 2004), 50. On Janet and Myers see also Taves, *Fits, Trances, and Visions*, 253–58.

30. As an extension of this, Taves would seem justified in calling the comparison between religious and nonreligious phenomena crucial for the construction of modern theories of religion. See Taves, "The Fragmentation of Consciousness," 63.

31. See Taylor, *Varieties of Religion Today*. This is not to say that this is Taylor's reason for selecting James, which he rather connects to his focus on individuality and personal religion. For Taylor the element of personal religion is what defines the arc of Western religious development and hence also lies at the root of secularism as diagnosed in his *A Secular Age*.

32. Taylor, *Varieties of Religion Today*, 63.

33. James, *Varieties of Religious Experience*: 139–56 (Lecture 8: "The Divided Self, and the Process of Unification"); 157–77 (Lecture 9: "Conversion"); 178–209 (Lecture 10: "Conversion—Concluded").

34. On the connection of the American psychologist Edwin Diller Starbuck (1866–1947) to James, see *Varieties*, 428.

35. The note in the edition has not located Emerson's quotation but refers instead to "Intellect" (*CW* 2:197): "Then, in a moment, and unannounced, the truth appears" (James, *Varieties*, 450). This citation does not capture the element of casualness, however, that I also see present in the idea of "hands off."

36. Compare this to what Emerson calls a succession of moods ("Experience," in *CW* 3:30), as these moods do not seem causally related. See above Chap. 1n55.

37. James, who never was a Methodist, may well have wanted to identify with Methodism as a stand-in for what would have been a more problematic identification with his father's Swedenborgianism. See Taves, *Fits, Trances, and Visions*, 270.

38. See Mark Ford in "Hardy's Neutral Tones, an LRB Podcast with Seamus Perry and Mark Ford," www.lrb.co.uk/2018/06/27/lrb-podcast/hardys-neutral-tones, at 25:20ff.

39. Thomas Hardy, "The Self-Unseeing," in *Poems of the Past and the Present* (New York: Harper and Brothers, 1902), 211–12.

40. On the poem's exploration of ambiguity and the paradox of consciousness see Peter Simpson, "The 'Self-Unseeing' and the Romantic Problem of Consciousness," *Victorian Poetry* 17, nos. 1–2 (1979): 49.

41. See Simpson, "The 'Self-Unseeing,'" 48. Simpson emphasizes how objects relay temporal meaning in the poem: the ancient floor, the former door.

42. See note 21 above.

43. On the fields of consciousness see also Richard R. Niebuhr, "William James on Religious Experience," in *The Cambridge Companion to William James*, ed. Ruth A. Putnam (Cambridge: Cambridge University Press, 1997), 225–28.

44. See Proudfoot, "William James on an Unseen Order," 55.

45. Odd as it may seem, in Emerson the impersonal seems comparable to the intimate in James. In the words of Sharon Cameron: "Thus in Emerson's account the impersonal enables the social world it appears to eradicate." See her "The Way of Life by Abandonment: Emerson's Impersonal," in *The Other Emerson*, ed. Branka Arsić and Cary Wolfe (Minneapolis: University of Minnesota Press, 2010), 9.

46. Lamberth, *William James*, 202. See also David C. Lamberth, "Interpreting the Universe after a Social Analogy: Intimacy, Panpsychism, and a Finite God in a Pluralistic Universe," in *The Cambridge Companion to William James*, ed. Ruth A. Putnam (Cambridge: Cambridge University Press, 1997), 237–59.

CONCLUSION

1. See on this development and all that it entails, starting from the twelfth century, Richard W. Southern, *Scholastic Humanism and the Unification of Europe*, vol. 1, *Foundations* (Oxford: Blackwell, 1995); and vol. 2, *The Heroic Age* (Oxford: Blackwell, 2001). Southern mentions that all subjects required a broadly similar progression from texts to individual problems to general systematic knowledge (1:11) and holds that the scholastic system held sway until the secularization of the twentieth century (1:13).

2. That retrieval was the subject of my book *From Paradise to Paradigm: A Study of Twelfth-Century Humanism* (Leiden: Brill, 2004).

3. Whereas angels are hierarchically placed above humans given their spiritual bodies, they are not created in the image of God and thus occupy a less privileged place.

4. It is more than a little confusing, and continues to some extent Maximus's invisibility in the West, that Denys Turner recognizes that Gregory of Nyssa and Maximus tower above Dionysius in the East but still privileges the latter as more influential in the West. See Denys Turner, *The Darkness of God: Negativity in Christian Mysticism* (Cambridge: Cambridge University Press, 1995), 11–18.

5. Their similarity is partly explained by their common roots in the allegorization of Genesis that originated with Philo.

6. I want to mention here again Virginia Burrus, *"Begotten, Not Made": Conceiving Manhood in Late Antiquity* (Stanford: Stanford University Press, 2000), which deals with the male-centrism of post-Nicene theology; and the recent dissertation by Bryce E. Rich, "Beyond Male and Female: Gender Essentialism and Orthodoxy"

254 *Notes to Conclusion*

(University of Chicago Divinity School, 2017), which analyzes some of the same problems couched in terms of gender essentialism.

7. See M. B. Pranger, with a reference to Robert Musil's *Man without Qualities*, in "Monastic Violence," in *Violence, Identity, and Self-Determination*, ed. Hent de Vries and Samuel Weber (Stanford: Stanford University Press, 1997), 44–57.

8. See Henri de Lubac, *Medieval Exegesis*, vol. 1, *The Four Senses of Scripture*, trans. M. Sebanc (Grand Rapids, MI: Eerdmans, 1998), 27: "As a result, theological science and the explication of Scripture cannot but be one and the same thing. In its most profound and far-reaching sense this estimation of the situation remains true even to our own day. But in its stricter and more immediate sense, this idea flourished right to the eve of the thirteenth century."

9. On the relationship of various modern theologians to scripture, especially Schleiermacher's theology of correlation, see Hans W. Frei, *Types of Christian Theology*, ed. Georg Hunsinger and William C. Placher (New Haven, CT: Yale University Press, 1992), 57–69. Obviously, Karl Barth would be an exception here. But as argued by Frei, his prioritization of Christian self-description creates tensions with the demands of second-order language and the relationship of theology to philosophy.

10. See Martin Jay, *Songs of Experience: Modern American and European Variations on a Universal Theme* (Berkeley: University of California Press, 2005), 78–88 (on Schleiermacher).

11. On Schleiermacher as influenced by Romanticism and Pietism see Jay, *Songs of Experience*, 88–102.

12. By contrast, twelfth-century allegorists did.

13. See on this Jan Rohls, " 'Der Winckelmann der griechischen Philosophie'— Schleiermachers Platonismus im Kontext," in *200 Jahre "Reden über die Religion,"* ed. Ulrich Barth and Claus-Dieter Osthövener (Berlin: Walter de Gruyter, 2000), 467.

14. Johannes Zachhuber sees Schleiermacher and Schelling as inaugurating the historical turn in nineteenth-century theology; see J. Zachhuber, "The Historical Turn," in *The Oxford Handbook of Nineteenth-Century Christian Thought*, ed. Joel D. S. Rasmussen, Judith Wolfe, and J. Zachhuber (Oxford: Oxford University Press, 2017), 57–60.

15. Friedrich Schleiermacher, "Fifth Speech: On the Religions," in *On Religion: Speeches to Its Cultured Despisers*, trans. and ed. Richard Crouter (Cambridge: Cambridge University Press, 2015), 113, 115; *KGA* 1.2, 314, 316.

16. Schleiermacher, 115; *KGA* 1.2, 316.

17. See Paul E. Capetz, "Friedrich Schleiermacher on the Old Testament," *Harvard Theological Review* 102, no. 3 (2009): 297–325.

18. Jay, *Songs of Experience*, 95–96. Rosenzweig later renounced subjectivist *Erlebnis* in favor of *Erfahrung*, a paradoxical notion of experience without a subject, comparable to Heidegger and Benjamin; see Jay, *Songs of Experience*, 129.

19. See Wayne Proudfoot, "From Theology to a Science of Religions: Jonathan

Edwards and William James on Religious Affections," *Harvard Theological Review* 82, no. 2 (1989): 149–68.

20. James receives no analysis in Oliver D. Crisp and Douglas A. Sweeney, eds., *After Jonathan Edwards: The Courses of the New England Theology* (Oxford: Oxford University Press, 2012). Representing an older, more humanistic tradition that includes Edwards with Emerson and James is William A. Clebsch, *American Religious Thought: A History* (Chicago: University of Chicago Press, 1973).

21. Another case of such a sudden shift of perspective occurs in "Experience" (*CW* 3:29): "Nature does not like to be observed, and likes that we should be her fools and playmates. We may have the sphere for our cricket-ball, but not a berry for our philosophy."

22. In Revelation 8:1, after the opening of the seventh seal, there is half an hour of silence, where John sees the angels around the throne and plagues are sent down. The "discomfit of the conclusions of nations" seems to uncover an apocalyptic tendency here.

23. See the critical apparatus in *CW* 3:190.

24. Cf. "Experience," *CW* 3:42: "Our life seems not present, so much as prospective."

25. Here Emerson is more radically engaged or perhaps more intimate than James, who does not "report" but is ever the "recorder."

Bibliography

Adams, Robert M. "Faith and Religious Knowledge." In Mariña, *Cambridge Companion to Friedrich Schleiermacher*, 35–51.

———. "Schleiermacher on Evil." *Faith and Philosophy* 13, no. 4 (1996): 563–83.

Albanese, Catherine L. *A Republic of Mind and Spirit: A Cultural History of American Metaphysical Religion*. New Haven, CT: Yale University Press, 2007.

Albrecht, James M. *Reconstructing Individualism: A Pragmatic Tradition from Emerson to Ellison*. New York: Fordham University Press, 2012.

Allen, Pauline, and Bronwen Neil, eds. *The Oxford Handbook of Maximus the Confessor*. Oxford: Oxford University Press, 2015.

Armstrong, Arthur Hilary, ed. *The Cambridge History of Later Greek and Early Medieval Philosophy*. Cambridge: Cambridge University Press, 1967.

Arsić, Branka. "Brain Walks: Emerson on Thinking." In Arsić and Wolfe, *The Other Emerson*, 59–97.

———. *On Leaving: A Reading in Emerson*. Cambridge, MA: Harvard University Press, 2010.

Arsić, Branka, and Cary Wolfe, eds. *The Other Emerson*. Minneapolis: University of Minnesota Press, 2010.

Athanasius. *On the Incarnation: The Treatise "De Incarnatione Verbi Dei."* Translated and edited by A Religious of C.S.M.V. With an introduction by C. S. Lewis. New York: St. Vladimir's Seminary Press, 1977.

Auerbach, Erich. *Mimesis: The Representation of Reality in Western Literature*. Translated by Willard R. Trask. Princeton, NJ: Princeton University Press, 1968.

Augustinus, Aurelius [Augustine]. *Concerning the City of God against the Pagans*. Translated by H. Bettenson. London: Penguin, 1972.

———. *Confessions*. Edited by James J. O'Donnell. 3 vols. Oxford: Oxford University Press, 1992.

———. *Confessions: A New Translation by Henry Chadwick*. Oxford: Oxford University Press, 1991.

———. *De ciuitate Dei, libri I–XXII*. Edited by B. Dombart and A. Kalb. Corpus Christianorum Series Latina [CCSL] 47 (1–10); 48 (11–22). Turnhout: Brepols, 1955.

———. *La Genèse au sens littéral en douze livres: De Genesi ad litteram libri duodecim.* Traduction, introduction et notes par P. Agaësse et A. Solignac. Vol. 1 (bks. 1–7); Vol. 2 (bks. 8–12). Œuvres de St. Augustin 48–49. Paris: Desclée de Brouwer, 1972.

———. *The Literal Meaning of Genesis.* In Augustine, *On Genesis.* Translated by Edmund Hill, O.P., 168–506. New York: New City Press, 2002.

———. *Retractationum libri duo.* Edited by Almit Mutzenbecher. Corpus Christianorum Series Latina [CCSL] 57. Turnhout: Brepols, 1984.

———. *Soliloquies and Immortality of the Soul.* Introduction, translation, and commentary by G. Watson. Warminster, UK: Aris and Phillips, 1990.

Balthasar, Hans Urs von. *Cosmic Liturgy: The Universe according to Maximus the Confessor.* Translated by Brian E. Daley. San Francisco: Ignatius Press, 2003. Based on the 3rd German edition, 1988.

Beierwaltes, W. "Duplex Theoria: Zu einer Denkform Eriugenas." In *Begriff und Metapher: Sprachform des Denkens bei Eriugena.* Edited by W. Beierwaltes, 39–64. Heidelberg: Carl Winter, 1990.

Bellah, Robert N. *Religion in Human Evolution: From the Paleolithic to the Axial Age.* Cambridge, MA: Belknap Press of Harvard University Press, 2011.

Berthold, George C., trans. *Maximus Confessor: Selected Writings.* Mahwah, NJ: Paulist Press, 1985.

Blowers, Paul M. *Drama of the Divine Economy: Creator and Creation in Early Christian Theology and Piety.* Oxford: Oxford University Press, 2012.

Boudignon, Christian, ed. *Maximi Confessoris Mystagogia: Una cum latina interpretatione Anastasii Bibliothecarii.* Turnhout: Brepols, 2001.

Bright, Pamela, ed. and trans. *Augustine and the Bible.* Notre Dame, IN: University of Notre Dame Press, 1999.

Brown, Peter. *Augustine of Hippo: A Biography.* London: Faber and Faber, 1967.

———. *The Body and Society: Men, Women, and Sexual Renunciation in Early Christianity.* New York: Columbia University Press, 1988.

Buell, Lawrence. *Emerson.* Cambridge, MA: Belknap Press of Harvard University Press, 2003.

———. *The Environmental Imagination: Thoreau, Nature Writing, and the Formation of American Culture.* Cambridge, MA: Belknap Press of Harvard University Press, 1995.

Burnyeat, Myles. "Idealism and Greek Philosophy: What Descartes Saw and Berkeley Missed." *Philosophical Review* 91, no. 1 (1982): 3–40.

Burrus, Virginia. *Ancient Christian Ecopoetics: Cosmologies, Saints, Things.* Philadelphia: University of Pennsylvania Press, 2019.

———. *"Begotten, Not Made": Conceiving Manhood in Late Antiquity.* Stanford: Stanford University Press, 2000.

Burrus, Virginia, Mark D. Jordan, and Karmen MacKendrick. *Seducing Augustine: Bodies, Desires, Confessions.* New York: Fordham University Press, 2010.

Busch, Emil W. *Burckhardt und Nietzsche im Revolutionszeitalter.* München: Fink, 2012.

Butler, Edward Cuthbert. *Western Mysticism: The Teaching of Saint Augustine, Gregory and Bernard on Contemplation and the Contemplative Life.* 1926. New York: Dover, 2003.

Cadava, Eduardo. *Emerson and the Climates of History.* Stanford: Stanford University Press, 1997.

Cameron, Michael. *Christ Meets Me Everywhere: Augustine's Early Figurative Exegesis.* Oxford: Oxford University Press, 2012.

Cameron, Sharon. "The Way of Life by Abandonment: Emerson's Impersonal." In Arsić and Wolfe, *The Other Emerson,* 3–40.

Capetz, Paul E. "Friedrich Schleiermacher on the Old Testament." *Harvard Theological Review* 102, no. 3 (2009): 297–325.

Cappuyns, Maïeul, O.S.B. *Jean Scot Érigène: Sa vie, son œuvre, sa pensée.* Bruxelles: Culture et Civilisation, 1969.

———. "Le '*De imagine*' de Grégoire de Nysse traduit par Jean Scot Érigène." *Recherches de théologie ancienne et médiévale* 32 (1965): 204–62.

Carbine, Rosemary P., and Hilda P. Koster, eds. *The Gift of Theology: The Contribution of Kathryn Tanner.* Minneapolis, MN: Fortress, 2015.

Carey, Phillip. *Inner Grace: Augustine in the Traditions of Plato and Paul.* Oxford: Oxford University Press, 2008.

Cattoi, Thomas. "Liturgy as Cosmic Transformation." In Allen and Neil, *Oxford Handbook of Maximus,* 414–35.

Cavell, Stanley. "Being Odd, Getting Even (Descartes, Emerson, Poe)." In *In Quest of the Ordinary,* 105–30. Reprinted in *Emerson's Transcendental Etudes,* 83–109.

———. *The Claim of Reason: Wittgenstein, Skepticism, Morality, and Tragedy.* Oxford: Oxford University Press, 1979.

———. *Conditions Handsome and Unhandsome: The Constitution of Emersonian Perfectionism.* Chicago: University of Chicago Press, 1990.

———. "Emerson, Coleridge, Kant (Term as Conditions)." In *In Quest of the Ordinary,* 27–49.

———. "An Emerson Mood." In *The Senses of Walden,* 152–60. Reprinted in *Emerson's Transcendental Etudes,* 20–32.

———. *Emerson's Transcendental Etudes.* Stanford: Stanford University Press, 2003.

———. *In Quest of the Ordinary: Lines of Skepticism and Romanticism.* Chicago: University of Chicago Press, 1994.

————. *The Senses of Walden: An Expanded Edition.* Chicago: University of Chicago Press, 1992.

————. "Thinking of Emerson." In *The Senses of Walden*, 123–38. Reprinted in *Emerson's Transcendental Etudes*, 10–19.

Chadwick, Henry. "Origen." In Armstrong, *Cambridge History*, 182–92.

Clark, Elizabeth A. "'Adam's Only Companion': Augustine and the Early Christian Debate on Marriage." *Recherches augustiniennes* 21 (1986): 139–62.

Clebsch, William A. *American Religious Thought: A History.* Chicago: University of Chicago Press, 1973.

Coakley, S., and Charles M. Stang, eds. *Re-thinking Dionysius the Areopagite.* Oxford: Wiley-Blackwell, 2009.

Colish, Marcia L. *The Mirror of Language: A Study in the Medieval Theory of Knowledge.* Lincoln: University of Nebraska Press, 1983.

Collingwood, R. G. *The Idea of Nature.* Oxford: Oxford University Press, 1945.

Cramer, Peter. *Baptism and Change in the Early Middle Ages, c. 200–c. 1150.* Cambridge: Cambridge University Press, 1993.

Crisp, Oliver D., and Douglas A. Sweeney, eds. *After Jonathan Edwards: The Courses of the New England Theology.* Oxford: Oxford University Press, 2012.

Crouter, Richard. Introduction to Friedrich Schleiermacher, *On Religion: Speeches to Its Cultured Despisers.* Translated and edited by Richard Crouter, xi–xxxix. Cambridge: Cambridge University Press, 2015.

Cullen, Christopher M. *Bonaventure.* Oxford: Oxford University Press, 2006.

Daley, Brian E. "Origen's *De principiis*: A Guide to the Principles of Christian Scriptural Interpretation." In *Nova et vetera: Patristic Studies in Honor of Thomas Patrick Halton*, edited by J. Petruccione, 3–21. Washington, DC: Catholic University of America Press, 1998.

DeHart, Paul. "f(S) I/s: The Instance of Pattern, or Kathryn Tanner's Trinitarianism." In Carbine and Koster, *The Gift of Theology*, 29–55.

De Mey, Peter. "The Defence of Revealed Christianity through an Appeal to Experience in Hume, Lessing, and Schleiermacher." American Academy of Religion, Schleiermacher Group, Denver 2001. www.academia.edu/14382595/The_Defence_of_Revealed_Christianity_through_an_Appeal_to_Experience_in_Hume_Lessing_and_Schleiermacher.

Dietrich, P., and D. F. Duclow. "Virgins in Paradise: Deification and Exegesis in Periphyseon V." In *Jean Scot Écrivain: Actes du IVe Colloque International Montréal, 28 août–2 septembre, 1983*, ed. Guy-H Allard, 29–49. Montreal: Bellarmin / Paris: Vrin, 1986.

Dillenberger, John, and Claude Welch. *Protestant Christianity Interpreted through Its Development.* 2nd ed. Upper Saddle River, NJ: Prentice Hall, 1998.

Dinsmore, Patrick D. *A Facing-Page Translation from German into English of Friedrich Schleiermacher's Kurze Darstellung des Spinozistischen Systems and Spinozismus*. Lewiston, NY: Edwin Mellen, 2013.

Dionysius the Areopagite. *Mystical Theology*. In *Pseudo-Dionysius: The Complete Works*. Translated by Colm Luibheid, 135–41. New York: Paulist Press, 1987.

Dole, Andrew C. *Schleiermacher on Religion and the Natural Order*. Oxford: Oxford University Press, 2010.

Dronke, Peter. *Fabula: Explorations into the Uses of Myth in Medieval Platonism*. Leiden: Brill, 1985.

Duclow, Donald F. "Nature as Speech and Book in John Scotus Eriugena." *Mediaevalia* 3 (1977): 131–40.

Dunham, Scott. *The Trinity and Creation in Augustine: An Ecological Analysis*. Albany: State University of New York Press, 2008.

Economou, George D. *The Goddess Natura in Medieval Literature*. Notre Dame, IN: Notre Dame University Press, 2002.

Elshtain, Jean Bethke. *Augustine and the Limits of Politics*. Notre Dame, IN: Notre Dame University Press, 1995.

Emerson, Ralph Waldo. "The American Scholar." In *CW* 1:52–70.

———. "Circles." In *CW* 2:177–90.

———. *The Collected Works of Ralph Waldo Emerson* [*CW*]. Edited by Alfred R. Ferguson et al. 10 vols. Cambridge, MA: Belknap Press of Harvard University Press, 1971–2013.

———. *The Collected Works of Ralph Waldo Emerson* [*CW*]. Vol. 1, *Nature, Addresses, and Lectures*. Edited by Alfred R. Ferguson. Cambridge, MA: Belknap Press of Harvard University Press, 1971.

———. *The Collected Works of Ralph Waldo Emerson* [*CW*]. Vol. 2, *Essays: First Series*. Edited by Alfred R. Ferguson and Jean Ferguson Carr. Cambridge, MA: Belknap Press of Harvard University Press, 1979.

———. *The Collected Works of Ralph Waldo Emerson* [*CW*]. Vol. 3, *Essays: Second Series*. Edited by Alfred R. Ferguson and Jean Ferguson Carr. Cambridge, MA: Belknap Press of Harvard University Press, 1983.

———. "Divinity School Address." In *CW* 1:76–93.

———. "Experience." In *CW* 3:27–49.

———. "Intellect." In *CW* 2:193–205.

———. "The Method of Nature." In *CW* 1:117–37.

———. *Nature*. In *CW* 1:7–45.

———. "Nature." In *CW* 3:97–114.

———. "Self-Reliance." In *CW* 2:27–51.

———. "The Transcendentalist." In *CW* 1:201–16.

Eriugena, Johannes Scottus. *Periphyseon.* Books 1–5. Edited by E. A. Jeauneau. Corpus Christianorum Continuatio Mediaevalis [CCCM] 161–65. Turnhout: Brepols, 1996–2003.

———. *Periphyseon.* Translated by I. P. Sheldon-Williams. Revised by J. J. O'Meara. Montreal/Washington: Bellarmin and Dumbarton Oaks, 1987.

Frei, Hans W. *The Eclipse of Biblical Narrative: A Study in Eighteenth and Nineteenth Century Hermeneutics.* New Haven, CT: Yale University Press, 1974.

———. *Types of Christian Theology.* Edited by Georg Hunsinger and William C. Placher. New Haven, CT: Yale University Press, 1992.

Frishman, J., W. Otten, and G. Rouwhorst, eds. *Religious Identity and the Problem of Historical Foundation: The Foundational Character of Authoritative Sources in the History of Christianity and Judaism.* Leiden: Brill, 2004.

Garrigues, Jean-Michel. *Maxime le Confesseur: La charité, avenir divin de l'homme.* Paris: Beauchesne, 1976.

Gersh, Stephen. "*Per se ipsum*: The Problem of Immediate and Mediate Causation in Eriugena and His Neoplatonic Predecessors." In *Jean Scot Érigène et l'histoire de la philosophie: Laon 7–12 juillet 1975,* edited by René Roques, 367–76. Paris: CNRS, 1977.

Gersh, Stephen, and Dermot Moran, eds. *Eriugena, Berkeley, and the Idealist Tradition.* Notre Dame, IN: Notre Dame Press, 2006.

Ghosh, Amitav. *The Great Derangement: Climate Change and the Unthinkable.* Chicago: University of Chicago Press, 2016.

Gilson, Étienne. "*Regio dissimilitudinis* de Platon à Saint Bernard de Clairvaux." *Mediaeval Studies* 9, no. 1 (1947): 108–30.

Godman, Peter. *The Silent Masters: Latin Literature and Its Censors in the High Middle Ages.* Princeton, NJ: Princeton University Press, 2001.

Gregory the Great. "Life of Benedict." *The Dialogues.* Book 2, chap. 24. In *Early Christian Lives,* edited by Caroline White, 165–204. London: Penguin, 1998.

Grégoire de Nazianze. *Discours 38–41.* Edited by C. Moreschini and P. Gallay. Paris: Cerf, 1990.

Grégoire de Nysse. *La création de l'homme.* Translated by Jean Laplace, S.J. Notes by Jean Daniélou, S.J. Paris: Cerf, 1944.

Gregorius Nyssenus. *De hominis opificio.* In *Patrologia Graeca* 44, edited by J. P. Migne, 123–256. Paris, 1863.

Gregory, Eric. *Politics and the Order of Love: An Augustinian Ethic of Democratic Citizenship.* Chicago: University of Chicago Press, 2008.

Grimm, Reinhold R. *Paradisus coelestis, paradisus terrestris: Zur Auslegungsgeschichte des Paradieses im Abendland bis um 1200.* Munich: Fink, 1977.

Guiu, Adrian N., ed. *A Companion to John Scottus Eriugena.* Leiden: Brill, 2020.

Hadot, Pierre. *The Veil of Isis: An Essay on the History of the Idea of Nature.* Cambridge, MA: Belknap Press of Harvard University Press, 2008.

Hardy, Thomas. "The Self-Unseeing." In *Poems of the Past and the Present.* New York: Harper and Brothers, 1902.

"Hardy's Neutral Tones, an LRB Podcast with Seamus Perry and Mark Ford." www .lrb.co.uk/2018/06/27/lrb-podcast/hardys-neutral-tones.

Harrison, Peter. *The Bible, Protestantism, and the Rise of Natural Science.* Cambridge: Cambridge University Press, 1998.

———. "Miracles, Early Modern Science, and Rational Religion." *Church History* 75, no. 3 (2006): 493–510.

———. *The Territories of Science and Religion.* Chicago: University of Chicago Press, 2015.

Harrison, Robert Pogue. *Forests: The Shadow of Civilization.* Chicago: University of Chicago Press, 1992.

———. *Gardens: An Essay on the Human Condition.* Chicago: University of Chicago Press, 2008.

Haskins, Henry Stanley. *Meditations in Wall Street.* Introduction by Albert Jay Nock. New York: William Morrow, 1940.

Heidegger, Martin. *Being and Time: A Translation of Sein und Zeit.* Translated by Joan Stambaugh. Revised and with a foreword by Dennis J. Schmidt. Albany: State University of New York Press, 2010.

Hellemans, B. S., W. Otten, and M. B. Pranger, eds. *On Religion and Memory.* New York: Fordham University Press, 2013.

Holopainen, Toivo J. *Dialectic and Theology in the Eleventh Century.* Leiden: Brill, 1996.

Jacobi, F. H. *Über die Lehre des Spinoza, in Briefen an den Herrn Moses Mendelssohn* [excerpts]. In Vallée, Lawson, and Chapple, *Spinoza Conversations,* 78–125.

James, William. "Is Radical Empiricism Solipsistic?" In *Essays in Radical Empiricism,* 119–22. Cambridge, MA: Harvard University Press, 1976.

———. "The Sentiment of Rationality." In *The Will to Believe,* 57–89.

———. *The Varieties of Religious Experience.* Cambridge, MA: Harvard University Press, 1985.

———. "The Will to Believe." In *The Will to Believe,* 13–33.

———. *The Will to Believe and Other Essays in Popular Philosophy.* Cambridge, MA: Harvard University Press, 1979.

Jay, Martin. *Songs of Experience: Modern American and European Variations on a Universal Theme.* Berkeley: University of California Press, 2005.

Jeauneau, Edouard. *Études Érigéniennes.* Paris: Études Augustiniennes, 1987.

———. "From Origen's *Periarchon* to Eriugena's *Periphyseon.*" In Otten and Allen, *Eriugena and Creation,* 139–82.

———. "La division des sexes chez Grégoire de Nysse et chez Jean Scot Érigène." In *Eriugena: Studien zu seinen Quellen*, edited by Werner Beierwaltes, 33–54. Heidelberg: Carl Winter, 1980. Reprinted in Jeauneau, *Études Érigéniennes*, 341–64.

———. "L'effort, le labeur." In Jeauneau, *Quatre thèmes érigéniens*, 47–59. Reprinted in Jeauneau, *Études Érigéniennes*, 243–55.

———. "Le symbolisme de la mer chez Jean Scot Érigène." In Jeauneau, *Études Érigéniennes*, 289–96.

———. "L'usage de la notion d'*integumentum* à travers les gloses de Guillaume de Conches." *Archives d'histoire doctrinale et littéraire du Moyen Age* 24 (1957): 35–100. Reprinted in E. Jeauneau, *Lectio philosophorum: Recherches sur l'école de Chartres*, 127–92. Amsterdam: Hakkert, 1973.

———. *Quatre thèmes érigéniens*. Montreal: Vrin, 1978.

Joas, Hans. *Die Macht des Heiligen: Eine Alternative zur Geschichte von der Entzauberung*. Berlin: Suhrkamp, 2017.

Justin Martyr. "The First Apology of Justin, the Martyr." In *Early Christian Fathers*, edited by Cyril C. Richardson, 225–89. New York: Touchstone, 1996.

Kannengiesser, Charles. "Athanasius of Alexandria and the Foundation of Traditional Christology." *Theological Studies* 34, no. 1 (1973): 103–13. Reprinted in Charles Kannengiesser, *Arius and Athanasius: Two Alexandrian Theologians*. Hampshire: Variorum, 1991.

———. *Handbook of Patristic Exegesis*. 2 vols. Leiden: Brill, 2004.

Kendig, Elizabeth. "La forme dialogique dans le *Periphyseon*: Recréer l'esprit." *Les études philosophiques*, no. 1 (Jan. 2013): 101–19.

Kirwan, Christopher. *Augustine*. London: Routledge, 1991.

Knuuttila, Simo. "Time and Creation in Augustine." In *The Cambridge Companion to Augustine*, edited by E. Stumpf and N. Kretzmann, 103–15. Cambridge: Cambridge University Press, 2001.

Kolakowski, Leszek. *God Owes Us Nothing: A Brief Remark on Pascal's Religion and on the Spirit of Jansenism*. Chicago: University of Chicago Press, 1995.

Ladner, Gerhart B. *The Idea of Reform: Its Impact on Christian Thought and Action in the Age of the Fathers*. Rev. ed. New York: Harper and Row, 1967.

Lamberth, David C. "Interpreting the Universe after a Social Analogy: Intimacy, Panpsychism, and a Finite God in a Pluralistic Universe." In Putnam, *Cambridge Companion to William James*, 237–59.

———. *William James and the Metaphysics of Experience*. Cambridge: Cambridge University Press, 1999.

Lamm, Julia A. *The Living God: Schleiermacher's Theological Appropriation of Spinoza*. University Park: Pennsylvania State University Press, 1996.

Latour, Bruno. *The Politics of Nature: How to Bring the Sciences into Democracy.* Cambridge, MA: Harvard University Press, 2004.

Lim, Richard. "The Politics of Interpretation in Basil of Caesarea's *Hexaemeron.*" *Vigiliae Christianae* 44, no. 4 (1990): 315–70.

Lloyd-Sidle, Elena. "A Thematic Introduction to and Outline of the *Periphyseon,* for the Alumnus." In Guiu, *Companion to John Scottus Eriugena,* 113–33.

Lollar, Joshua. "Reception of Maximian Thought in the Modern Era." In Allen and Neil, *Oxford Handbook of Maximus,* 564–80.

———. *To See into the Life of Things: The Contemplation of Nature in Maximus the Confessor and His Predecessors.* Turnhout: Brepols, 2013.

Louth, Andrew. *Maximus the Confessor.* London: Routledge, 1996.

Lubac, Henri de. *History and Spirit: The Understanding of Scripture according to Origen.* 1950. San Francisco: Ignatius, 2007.

———. *Medieval Exegesis.* Vol. 1, *The Four Senses of Scripture.* Translated by M. Sebanc. Grand Rapids, MI: Eerdmans, 1998.

Lyotard, Jean-François. *The Confession of Augustine.* Translated by Richard Beardsworth. Stanford: Stanford University Press, 2000.

MacCormack, Sabine. "Augustine Reads Genesis." *Augustinian Studies* 39, no. 1 (2008): 5–47.

Mander, William. "Pantheism." *Stanford Encyclopedia of Philosophy Archive.* Winter 2016 Edition. https://plato.stanford.edu/archives/win2016/entries/pantheism.

Mariña, Jacqueline, ed. *The Cambridge Companion to Friedrich Schleiermacher.* Cambridge: Cambridge University Press, 2005.

———. "Christology and Anthropology in Friedrich Schleiermacher." In Mariña, *Cambridge Companion to Friedrich Schleiermacher,* 151–70.

Marion, Jean-Luc. *Au lieu de soi: L'approche de Saint Augustin.* Paris: PUF, 2008.

———. *In the Self's Place: The Approach of Saint Augustine.* Translated by Jeffrey L. Kosky. Stanford: Stanford University Press, 2012.

———. "Resting, Moving, Loving: The Access to the Self according to Saint Augustine." In Otten, "The Augustinian Moment," 24–42.

———. "A Saturated Phenomenon." *Filozofia* 62, no. 5 (2007): 378–402.

Markus, Robert A. *Christianity and the Secular.* Notre Dame, IN: University of Notre Dame Press, 2006.

———. *The End of Ancient Christianity.* Cambridge: Cambridge University Press, 1990.

———. *Signs and Meanings: World and Text in Ancient Christianity.* Eugene, OR: Wipf and Stock, 2011.

Maximos the Confessor. *Ambiguum 7.* In *On Difficulties,* 1:75–141.

———. *Ambiguum 7* (translation). In Blowers and Wilken, *Cosmic Mystery of Jesus Christ,* 45–74.

———. *Ambiguum* 10. In *On Difficulties*, 1:151–343.

———. *Ambiguum* 10 (translation). In Louth, *Maximus the Confessor*, 94–154.

———. *Ambiguum* 31. In *On Difficulties*, 2:38–52.

———. *Ambiguum* 33. In *On Difficulties*, 2:63–65.

———. *Ambiguum* 41. In *On Difficulties*, 2:103–21.

———. *Ambiguum* 41 (partial translation). In Louth, *Maximus the Confessor*, 156–62.

———. *On Difficulties in the Church Fathers: The "Ambigua."* Edited by Nicholas Constas. 2 vols. Cambridge, MA: Harvard University Press, 2014.

Maximus Confessor. "The Church's Mystagogy." In Berthold, *Maximus Confessor: Selected Writings*, 181–225.

———. *On the Cosmic Mystery of Jesus Christ: Selected Writings from Maximus the Confessor*. Translated by P. M. Blowers and R. L. Wilken. New York: St. Vladimir's Seminary Press, 2003.

McGinn, Bernard. *The Foundations of Mysticism: Origins to the Fifth Century*. Vol. 1 of *The Presence of God*. New York: Crossroad, 1992.

———. *The Growth of Mysticism: Gregory the Great through the 12th Century*. Vol. 2 of *The Presence of God*. New York: Crossroad, 1994.

———. "The Human Person as Image of God: I. Western Christianity." In McGinn and Meyendorff, *Christian Spirituality*, 312–30.

McGinn, Bernard, and John Meyendorff, eds. *Christian Spirituality: Origins to the Twelfth Century*. New York: Crossroad, 1988.

McGrath, Alistair. *Emil Brunner: A Reappraisal*. Oxford: Wiley Blackwell, 2014.

Meszaros, Andrew. *The Prophetic Church: History and Doctrinal Development in John Henry Newman and Yves Congar*. Oxford: Oxford University Press, 2016.

Moran, Dermot. "Idealism in Medieval Philosophy: The Case of Johannes Scottus Eriugena." *Medieval Philosophy and Theology* 8, no. 1 (1999): 53–82.

———. "Jean Scot Érigène, la connaissance de soi et la tradition idéaliste." *Les études philosophiques*, no. 1 (Jan. 2013): 29–56.

———. "Pantheism from John Scottus Eriugena to Nicholas of Cusa." *American Catholic Philosophical Quarterly* 64, no. 1 (1990): 131–52.

———. *The Philosophy of John Scottus Eriugena: A Study of Idealism in the Middle Ages*. Cambridge: Cambridge University Press, 1989.

———. "*Spiritualis incrassatio*: Eriugena's Intellectualist Immaterialism: Is It an Idealism?" In Gersh and Moran, *Eriugena*, 123–50.

Newman, Barbara. *God and the Goddesses: Vision, Poetry, and Belief in the Middle Ages*. Philadelphia: University of Pennsylvania Press, 2003.

Newman, John Henry. *An Essay on the Development of Christian Doctrine*. Cambridge: Cambridge University Press, 2010.

Niebuhr, Richard R. "William James on Religious Experience." In Putnam, *Cambridge Companion to William James*, 214–36.

Nietzsche, F. W. *The Complete Works of Friedrich Nietzsche*. Vol. 8, *Beyond Good and Evil/On the Genealogy of Morals*. Translated by Adrian Del Caro. Stanford: Stanford University Press, 2014.

———. *On the Advantage and Disadvantage of History for Life*. Translated by Peter Preuss. Indianapolis, IN: Hackett, 1980.

Nightingale, Andrea. *Once Out of Nature: Augustine on Time and the Body*. Chicago: University of Chicago Press, 2011.

O'Connell, M. J., trans. *The Byzantine Liturgy: Symbolic Structure and Faith Expression*. New York: Pueblo, 1986.

O'Meara, John J. *Eriugena*. Oxford: Clarendon, 1988.

Otten, Willemien. *The Anthropology of Johannes Scottus Eriugena*. Leiden: Brill, 1991.

———. "Anthropology between *Imago Mundi* and *Imago Dei*: The Place of Johannes Scottus Eriugena in the Tradition of Christian Thought." In *Studia Patristica*. Vol. 43, *Augustine, Other Latin Writers*, edited by F. Young, M. Edwards, and P. Parvis, 459–72. Leuven: Peeters, 2006.

———. "Augustine on Marriage, Monasticism and the Community of the Church." *Theological Studies* 59 (1998): 385–405.

———, ed. "The Augustinian Moment." Special issue, *Journal of Religion* 91, no. 1 (2011).

———. "Christianity's Content: (Neo)Platonism in the Middle Ages, Its Theoretical and Theological Appeal." *Numen* 63, nos. 2–3 (2016): 245–70.

———. "Cosmos and Liturgy from Maximus to Hans Urs von Balthasar (with an excursion on H. J. Schulz)." In *Sanctifying Texts, Transforming Rituals: Encounters in Liturgical Studies*, edited by P. van Geest, M. Poorthuis, and H. E. G. Rose, 153–69. Leiden: Brill, 2017.

———. "Creation and Epiphanic Incarnation: Reflections on the Future of Natural Theology from an Eriugenian-Emersonian Perspective." In Hellemans, Otten, and Pranger, *On Religion and Memory*, 64–88.

———. "The Dialectic of the Return in Eriugena's *Periphyseon*." *Harvard Theological Review* 84, no. 4 (1991): 399–421.

———. "Does the Canon Need Converting? A Meditation on Augustine's *Soliloquies*, Eriugena's *Periphyseon*, and the Dialogue with the Religious Past." In *How the West Was Won: Essays on Literary Imagination, the Canon, and the Christian Middle Ages for Burcht Pranger*, edited by W. Otten, A. Vanderjagt, and H. de Vries, 195–223. Leiden: Brill, 2010.

———. "Eriugena, Emerson, and the Poetics of Universal Nature." In *Metaphysical Patterns in Platonism: Ancient, Medieval, Renaissance, and Modern Times*, edited by John F. Finamore and Robert M. Berchman, 147–63. New Orleans: University Press of the South, 2007. Reprint: Wiltshire, UK: Prometheus Trust, 2014, 123–37.

———. "Eriugena and Emerson on Nature and the Self." In Otten and Allen, *Eriugena and Creation*, 503–38.

———. "Eriugena on Natures (Created, Human and Divine): From Christian-Platonic Metaphysics to Early-Medieval Protreptic." In *Philosophie et théologie chez Jean Scot Érigène*, edited by Isabelle Moulin, 113–33. Paris: Vrin, 2016.

———. *From Paradise to Paradigm: A Study of Twelfth-Century Humanism.* Leiden: Brill, 2004.

———. "Ideals of Community in Late Antiquity: John Cassian and Gregory the Great on Communicating Sanctity." In *Seeing the Invisible in Late Antiquity and the Early Middle Ages: Papers from "Verbal and Pictorial Imaging: Representing and Accessing Experience of the Invisible: 400–1000,"* edited by G. de Nie, K. F. Morrison, H. L. Kessler, and M. Mostert, 121–39. Turnhout: Peeters, 2005.

———. "Le langage de l'union mystique: Le désir et le corps dans l'œuvre de Jean Scot Érigène et de Maître Eckhart." *Les études philosophiques*, no. 1 (Jan. 2013): 121–41.

———. "The Long Shadow of Human Sin: Augustine on Adam, Eve, and the Fall." In *Out of Paradise: Eve and Adam and Their Interpreters*, edited by B. E. J. H. Becking and S. A. Hennecke, 29–49. Sheffield: Sheffield Phoenix Press, 2010.

———. "Medieval Latin Humanism." *Encyclopedia of Mediterranean Humanism*, edited by Houari Touati, Spring 2016. www.encyclopedie-humanisme.com/?Medieval-Latin-Humanism.

———. "Nature and Scripture: Demise of a Medieval Analogy." *Harvard Theological Review* 88, no. 2 (1995): 257–84.

———. "Nature and the Representation of Divine Creation in the Twelfth Century." In *Communities of Learning: Networks and the Shaping of Intellectual Identity in Europe, 1100–1500*, edited by C. Mews and J. N. Crossley, 57–74. Turnhout: Brepols, 2011.

———. "Nature as a Theological Problem: An Emersonian Response to Lynn White." In *Responsibility and the Enhancement of Life: Essays in Honor of William Schweiker*, ed. Günter Thomas and Heike Springhart, 265–80. Leipzig: Evangelische Verlagsanstalt, 2017.

———. "Nature as Religious Force in Eriugena and Emerson." In *Religion: Beyond a Concept*, edited by Hent de Vries, 354–67. New York: Fordham University Press, 2008.

———. "On *Sacred Attunement*, Its Meaning and Consequences: A Meditation on Christian Theology." *Journal of Religion* 93, no. 4 (2013): 478–94.

———. "Realized Eschatology or Philosophical Idealism: The Case of Eriugena's *Periphyseon*." In *Ende und Vollendung: Eschatologische Perspektiven im Mittelalter*, edited by J. A. Aertsen and M. Pickavé, 373–87. New York: De Gruyter, 2002.

———. "Religion as *Exercitatio Mentis*: A Case for Theology as a Humanist Discipline." In *Christian Humanism: Essays Offered to Arjo Vanderjagt on the Occasion of his Sixtieth Birthday*, edited by Z. R. W. M. von Martels and A. MacDonald, 59–73. Leiden: Brill, 2009.

———. Review of *Au lieu de soi: L'approche de Saint Augustin*, by Jean-Luc Marion. *Continental Philosophy Review* 42, no. 4 (2010): 597–602.

———. "Suspended between Cosmology and Anthropology: *Natura*'s Bond in Eriugena's *Periphyseon*." In Guiu, *Companion to John Scottus Eriugena*, 189–212.

———. "The Tension between Word and Image in Christianity." In *Iconoclasm and Iconoclash: Struggle for Religious Identity*, edited by P. van Geest, D. Müller, W. van Asselt, and Th. Salemink, 33–48. Leiden: Brill, 2007.

———. "West and East: Prayer and Cosmos in Augustine and Maximus Confessor." In *The Early Christian Mystagogy of Prayer*, edited by P. van Geest, H. van Loon, and G. de Nie, 319–37. Leuven: Peeters, 2019.

Otten, W., and M. I. Allen, eds. *Eriugena and Creation*. Proceedings of the Eleventh International Conference on Eriugenian Studies, held in honor of Edouard Jeauneau, Chicago, 9–12 Nov. 2011. Turnhout: Brepols, 2014.

Pals, Daniel L. *Seven Theories of Religion*. New York: Oxford University Press, 1996.

Pannenberg, Wolfhart. *The Historicity of Nature: Essays on Science and Theology*. Edited by Niels Henrik Gregersen. West Conshohocken, PA: Templeton Foundation Press, 2008.

Pollmann, Karla. "Augustine, Genesis, and Controversy." *Augustinian Studies* 38, no. 1 (2007): 203–16.

Pollmann, Karla, and W. Otten, eds. *The Oxford Guide to the Historical Reception of Augustine*. 3 vols. Oxford: Oxford University Press, 2013.

Pranger, M. B. *Eternity's Ennui: Temporality, Perseverance and Voice in Augustine and Western Literature*. Leiden: Brill, 2010.

———. "Monastic Violence." In *Violence, Identity, and Self-Determination*, edited by Hent de Vries and Samuel Weber, 44–57. Stanford: Stanford University Press, 1997.

———. Review of *Once Out of Nature*, by Andrea Nightingale. *Journal of Religion* 92, no. 3 (2012): 423–24.

Proudfoot, Wayne. "From Theology to a Science of Religions: Jonathan Edwards and William James on Religious Affections." *Harvard Theological Review* 82, no. 2 (1989): 149–68.

———. "Pragmatism and 'An Unseen Order' in *Varieties*." In *William James and a Science of Religions*, 31–47.

———. *Religious Experience.* Berkeley: University of California Press, 1985.

———, ed. *William James and a Science of Religions: Reexperiencing "The Varieties of Religious Experience."* New York: Columbia University Press, 2004.

———. "William James on an Unseen Order." *Harvard Theological Review* 93, no. 1 (2000): 51–66.

Putnam, Ruth A., ed. *The Cambridge Companion to William James.* Cambridge: Cambridge University Press, 1997.

Ramelli, Ilaria L. E. *The Christian Doctrine of Apokatastasis: A Critical Assessment from the New Testament to Eriugena.* Leiden: Brill, 2013.

Rich, Bryce E. "Beyond Male and Female: Gender Essentialism and Orthodoxy." PhD diss., University of Chicago Divinity School, 2017.

Richardson, Robert D., Jr. *Emerson: The Mind on Fire.* Berkeley: University of California Press, 1995.

———. "Emerson and Nature." In *The Cambridge Companion to Ralph Waldo Emerson,* edited by Joel Porte and Saundra Morris, 97–105. Cambridge: Cambridge University Press, 1999.

———. "Schleiermacher and the Transcendentalists." In *Transient and Permanent: The Transcendentalist Movement and Its Contexts,* edited by Charles Capper and Conrad E. Wright, 121–47. Boston: Massachusetts Historical Society, 1999.

Ricoeur, Paul. *Time and Narrative.* 3 vols. Translated by Kathleen MacLaughlin and David Pellauer. Chicago: University of Chicago Press, 1984–88.

Rist, John M. *Augustine: Ancient Thought Baptized.* Cambridge: Cambridge University Press, 1994.

Robertson, Kellie. *Nature Speaks: Medieval Literature and Aristotelian Philosophy.* Philadelphia: University of Pennsylvania Press, 2017.

Rogers, Eugene F., Jr. "Tanner's Non-competitive Account and the Blood of Christ: Where Eucharistic Theology Meets the Evolution of Ritual." In Carbine and Koster, *The Gift of Theology,* 139–57.

Rohls, Jan. "'Der Winckelmann der griechischen Philosophie'—Schleiermachers Platonismus im Kontext." In *200 Jahre "Reden über die Religion." Akten des 1. Internationalen Kongresses der Schleiermacher-Gesellschaft Halle, 14.–17. März 1999,* edited by Ulrich Barth and Claus-Dieter Osthövener, 467–96. Berlin: Walter de Gruyter, 2000.

Rorty, Richard. "Some Inconsistencies in James's *Varieties.*" In Proudfoot, *William James and a Science of Religions,* 86–97.

Rousseau, Philip. "Human Nature and Its Material Setting in Basil of Caesarea's Sermons on the Creation." *Heythrop Journal* 49 (2008): 222–39.

Rouwhorst, Gerard. "Historical Periods as Normative Sources: The Appeal to the Past in the Research on Liturgical History." In Frishman, Otten, and Rouwhorst, *Religious Identity,* 494–512.

Rubenstein, Mary-Jane. *Pantheologies: Gods, Worlds, Monsters.* New York: Columbia University Press, 2018.

Scheck, Thomas P., trans. *St. Jerome: Commentary on Matthew.* Washington, DC: Catholic University of America Press, 2008.

Schleiermacher, Friedrich. *Der christliche Glaube nach den Grundsätzen der evangelischen Kirche im Zusammenhange dargestellt.* Zweite Auflage (1830/31). Edited by Rolf Schäfer. Berlin: De Gruyter, 2008.

———. "Fifth Speech: On the Religions." In *On Religion,* 95–124.

———. "First Speech: Apology." In *On Religion,* 3–17.

———. *On Religion: Speeches to Its Cultured Despisers.* Translated and edited by Richard Crouter. Cambridge: Cambridge University Press, 2015.

———. "Second Speech: On the Essence of Religion." In *On Religion,* 18–54.

———. *Über die Religion: Reden an die Gebildeten unter ihren Verächtern (1799).* Edited by Günter Meckenstock. Berlin: Walter de Gruyter, 2001.

Schulz, Hans-Joachim. *Die byzantinische Liturgie: Glaubenszeugnis und Symbolgestalt.* Trier: Paulinus, 2000.

———. *Die byzantinische Liturgie: Vom Werden ihrer Symbolgestalt.* Freiburg: Lambertus, 1964.

Sheehan, Jonathan. *The Enlightenment Bible: Translation, Scholarship, Culture.* Princeton, NJ: Princeton University Press, 2005.

Sheldon-Williams, I. P. "St. Maximus the Confessor." In Armstrong, *Cambridge History,* 492–517.

Siewers, Alfred K. *Strange Beauty: Ecocritical Approaches to Early Medieval Landscape.* New York: Palgrave MacMillan, 2009.

Simpson, Peter. "The 'Self-Unseeing' and the Romantic Problem of Consciousness." *Victorian Poetry* 17, nos. 1–2 (1979): 45–50.

Slater, Michael R. *William James on Ethics and Faith.* Cambridge: Cambridge University Press, 2009.

Smith, J. Warren. *Passion and Paradise: Human and Divine Emotion in the Thought of Gregory of Nyssa.* New York: Herder-Crossroad, 2004.

Southern, Richard W. *Scholastic Humanism and the Unification of Europe.* Vol. 1, *Foundations.* Oxford: Blackwell, 1995; Vol. 2, *The Heroic Age.* Oxford: Blackwell, 2001.

Speer, A. *Die entdeckte Natur: Untersuchungen zu Begründungsversuchen einer "scientia naturalis" im 12. Jahrhundert.* Leiden: Brill, 1995.

Spinoza, Benedict de. *Ethics.* Edited and translated by Edwin Curley, with an introduction by Stuart Hampshire. London: Penguin, 1996.

Stack, George J. *Nietzsche and Emerson: An Elective Affinity.* Athens: Ohio University Press, 1992.

Stang, Charles M. *Apophasis and Pseudonymity in Dionysius the Areopagite: "No Longer I."* Oxford: Oxford University Press, 2012.

Stock, Brian. *Augustine's Inner Dialogue: The Philosophical Soliloquy in Late Antiquity*. Cambridge: Cambridge University Press, 2010.

Tanner, Kathryn. *God and Creation in Christian Theology: Tyranny or Empowerment?* Minneapolis, MN: Fortress, 2005.

———. "Human Freedom, Human Sin, and God the Creator." In *The God Who Acts: Philosophical and Theological Explorations*, edited by Thomas F. Tracy, 111–35. University Park: Pennsylvania State University Press, 2005.

Tanner, Norman P., S.J., ed. and trans. *Decrees of the Ecumenical Councils*. Vol. 1, *Nicaea I to Lateran V*. London/Washington: Sheed and Ward/Georgetown University Press, 1990.

Taves, Ann. *Fits, Trances, and Visions: Experiencing Religion and Explaining Experience from Wesley to James*. Princeton, NJ: Princeton University Press, 1999.

———. "The Fragmentation of Consciousness and *The Variety of Religious Experience*: William James's Contribution to a Theory of Religion." In Proudfoot, *William James and a Science of Religions*, 48–72.

———. *Religious Experience Reconsidered: A Building-Block Approach to the Study of Religion and Other Special Things*. Princeton, NJ: Princeton University Press, 2009.

Taylor, Charles. *A Secular Age*. Cambridge, MA: Belknap Press of Harvard University Press, 2007.

———. *Varieties of Religion Today: William James Revisited*. Cambridge, MA: Harvard University Press, 2002.

Thunberg, Lars. "The Human Person as Image of God: I. Eastern Christianity." In McGinn and Meyendorff, *Christian Spirituality*, 291–312.

———. *Microcosm and Mediator: The Theological Anthropology of Maximus Confessor*. 1965. Chicago: Open Court Press, 1995.

Tilley, Terrence W. *Evils of Theodicy*. Eugene, OR: Wipf and Stock, 2000.

Tollefsen, Torstein T. "Christocentric Cosmology." In Allen and Neil, *Oxford Handbook of Maximus*, 307–21.

Tracy, David. *Plurality and Ambiguity: Hermeneutics, Religion, Hope*. Chicago: University of Chicago Press, 1987.

Turner, Denys. *The Darkness of God: Negativity in Christian Mysticism*. Cambridge: Cambridge University Press, 1995.

Vallée, Gérard. *The Spinoza Conversations between Lessing and Jacobi: Texts with Excerpts from the Ensuing Controversy*. Introduced by Gérard Vallée. Translated by G. Vallée, J. B. Lawson, and C. G. Chapple. Lanham, MD: University Press of America, 1988.

Vanderjagt, Arjo. "Political Theology." In Pollmann and Otten, *Oxford Guide*, 3:1562–69.

Vannier, Marie-Anne. *"Creatio," "Conversio," "Formatio" chez S. Augustin* (Fribourg: Éditions universitaires, 1997).

———. "Le rôle de l'hexaéméron dans l'interprétation augustinienne de la création." *Revue des sciences philosophiques et théologiques* 71 (1987): 537–47.

Vessey, Mark, ed. *A Companion to Augustine.* Oxford: Wiley-Blackwell, 2012.

Watson, Robert N. *Back to Nature: The Green and the Real in the Late Renaissance.* Philadelphia: University of Pennsylvania Press, 2006.

Wetzel, James. *Augustine: A Guide for the Perplexed.* New York: Continuum, 2010.

———. Review of *Once Out of Nature*, by Andrea Nightingale. *Notre Dame Philosophical Reviews: An Electronic Journal*, Jan. 4, 2012. http://ndpr.nd.edu/news/once-out-of-nature-augustine-on-time-and-the-body.

White, Lynn, Jr. "The Historical Roots of Our Ecological Crisis." *Science*, n.s., 155, no. 3767 (March 10, 1967): 1203–7.

Wilson, David S. *Darwin's Cathedral: Evolution, Religion, and the Nature of Society.* Chicago: University of Chicago Press, 2003.

Wirzba, Norman. *From Nature to Creation: A Christian Vision for Understanding and Loving Our World.* Grand Rapids, MI: Baker Academic, 2015.

———. *The Paradise of God: Renewing Religion in an Ecological Age.* Oxford: Oxford University Press, 2003.

Wulf, Andrea. *The Invention of Nature: Alexander von Humboldt's New World.* New York: Vintage, 2015.

Yeats, William Butler. "Sailing to Byzantium." 1928. In *The Collected Poems of W. B. Yeats.* Edited by Richard Finneran. New York: Simon and Schuster, 1989.

Zachhuber, Johannes. "The Historical Turn." In *The Oxford Handbook of Nineteenth-Century Christian Thought*, edited by Joel D. S. Rasmussen, Judith Wolfe, and Johannes Zachhuber, 53–71. Oxford: Oxford University Press, 2017.

Index

Abelard, Peter, 35
Adams, Robert M., 154–57, 249
aesthetics, 4, 167–68, 173, 207
Alan of Lille, 18, 20, 33, 228; *Plaint of Nature,* 18, 33
Albert the Great, 156
allegory, 3, 8, 18–21, 37, 39, 42, 79, 92, 94, 103, 116, 202, 204, 224, 253–54. *See also under* scripture
Alline, Henry, 184
Amalric of Bene, 131
Ambrose, 39
angels, 42, 54–56, 78, 85, 110, 118, 121–24, 215, 231, 253, 255
Anglicanism, 184
annihilation, 62, 75
Anselm of Canterbury, 40, 229
anthropology, 8, 16, 33, 39, 52, 59–66, 70–74, 104, 113, 117, 137, 147, 150, 161, 163, 177–78, 204, 208–11, 214, 221–22, 225, 231, 241–45, 248
apocalypse, 23, 123, 215, 255
apocatastasis, 57
apology, 137–38, 144–47, 151, 156, 167, 207, 209, 212, 247, 250. *See also under* Schleiermacher
apophasis. See negative theology
Aquinas, Thomas, 2–3, 154, 156, 205, 245; *Summa Theologiae,* 3, 156, 205

Aristotelian categories, 3, 89, 116, 134; Aristotelianism, 41, 89, 131, 134, 166, 224–25; God as prime mover, 41, 83
Arsić, Branka, 22, 220, 226, 228, 253
ascent, 41, 50, 53, 56–58, 62–63, 70
asceticism, 40, 43, 51, 68, 71–72, 75–83, 98, 100–3, 116, 121, 184, 199, 201, 233, 237, 239
Athanasius, 57, 236–37
atheism, 134, 136
atonement, 26, 174
Auerbach, Erich, 120, 245
Augustine of Hippo, creation as *omnia simul,* 87, 96–97, 103, 107–8, 242; *distentio animi,* 81, 98, 239; divine discourse (*diuina eloquentia*), 112, 115–16, 120; eloquence of things (*eloquentia rerum*), 112–16, 119–21, 244; works: *Confessions,* 5, 37, 73, 84–86, 90, 93–94, 97, 103–6, 109, 114, 176, 237–39; *De bono coniugali,* 80, 100, 243; *De ciuitate dei (The City of God),* 80, 231, 243; *De Genesi ad litteram,* 86, 90–95, 103–6, 112, 240–43; *De trinitate dei,* 85; *De doctrina christiana (On Christian Teaching),* 111–13; *Retractationes,* 86, 93, 96, 107, 240; *Soliloquies,* 85, 231, 239, 242
automatism, 185, 188, 190

The index has been compiled by Lauren Beversluis.

Cultural Memory | *in the Present*

Theodore W. Jennings, Jr., *Outlaw Justice: The Messianic Politics of Paul*
Alexander Etkind, *Warped Mourning: Stories of the Undead in the
    Land of the Unburied*
Denis Guénoun, *About Europe: Philosophical Hypotheses*
Maria Boletsi, *Barbarism and Its Discontents*
Sigrid Weigel, *Walter Benjamin: Images, the Creaturely, and the Holy*
Roberto Esposito, *Living Thought: The Origins and Actuality of Italian Philosophy*
Henri Atlan, *The Sparks of Randomness, Volume 2: The Atheism of Scripture*
Rüdiger Campe, *The Game of Probability: Literature and Calculation
    from Pascal to Kleist*
Niklas Luhmann, *A Systems Theory of Religion*
Jean-Luc Marion, *In the Self's Place: The Approach of Saint Augustine*
Rodolphe Gasché, *Georges Bataille: Phenomenology and Phantasmatology*
Niklas Luhmann, *Theory of Society, Volume 1*
Alessia Ricciardi, *After* La Dolce Vita: *A Cultural Prehistory of Berlusconi's Italy*
Daniel Innerarity, *The Future and Its Enemies: In Defense of Political Hope*
Patricia Pisters, *The Neuro-Image: A Deleuzian Film-Philosophy of
    Digital Screen Culture*
François-David Sebbah, *Testing the Limit: Derrida, Henry, Levinas, and the
    Phenomenological Tradition*
Erik Peterson, *Theological Tractates*, edited by Michael J. Hollerich
Feisal G. Mohamed, *Milton and the Post-Secular Present: Ethics, Politics, Terrorism*
Pierre Hadot, *The Present Alone Is Our Happiness, Second Edition: Conversations
    with Jeannie Carlier and Arnold I. Davidson*
Yasco Horsman, *Theaters of Justice: Judging, Staging, and Working Through in
    Arendt, Brecht, and Delbo*
Jacques Derrida, *Parages*, edited by John P. Leavey
Henri Atlan, *The Sparks of Randomness, Volume 1: Spermatic Knowledge*
Rebecca Comay, *Mourning Sickness: Hegel and the French Revolution*
Djelal Kadir, *Memos from the Besieged City: Lifelines for Cultural Sustainability*
Stanley Cavell, *Little Did I Know: Excerpts from Memory*
Jeffrey Mehlman, *Adventures in the French Trade: Fragments Toward a Life*
Jacob Rogozinski, *The Ego and the Flesh: An Introduction to Egoanalysis*
Marcel Hénaff, *The Price of Truth: Gift, Money, and Philosophy*
Paul Patton, *Deleuzian Concepts: Philosophy, Colonialization, Politics*
Michael Fagenblat, *A Covenant of Creatures: Levinas's Philosophy of Judaism*
Stefanos Geroulanos, *An Atheism That Is Not Humanist Emerges
    in French Thought*
Andrew Herscher, *Violence Taking Place: The Architecture of the Kosovo Conflict*
Hans-Jörg Rheinberger, *On Historicizing Epistemology: An Essay*
Jacob Taubes, *From Cult to Culture*, edited by Charlotte Fonrobert
    and Amir Engel

Jacques Derrida and Elisabeth Roudinesco, *For What Tomorrow . . . : A Dialogue*

Elisabeth Weber, *Questioning Judaism: Interviews by Elisabeth Weber*

Jacques Derrida and Catherine Malabou, *Counterpath: Traveling with Jacques Derrida*

Martin Seel, *Aesthetics of Appearing*

Nanette Salomon, *Shifting Priorities: Gender and Genre in Seventeenth-Century Dutch Painting*

Jacob Taubes, *The Political Theology of Paul*

Jean-Luc Marion, *The Crossing of the Visible*

Eric Michaud, *The Cult of Art in Nazi Germany*

Anne Freadman, *The Machinery of Talk: Charles Peirce and the Sign Hypothesis*

Stanley Cavell, *Emerson's Transcendental Etudes*

Stuart McLean, *The Event and Its Terrors: Ireland, Famine, Modernity*

Beate Rössler, ed., *Privacies: Philosophical Evaluations*

Bernard Faure, *Double Exposure: Cutting Across Buddhist and Western Discourses*

Alessia Ricciardi, *The Ends of Mourning: Psychoanalysis, Literature, Film*

Alain Badiou, *Saint Paul: The Foundation of Universalism*

Gil Anidjar, *The Jew, the Arab: A History of the Enemy*

Jonathan Culler and Kevin Lamb, eds., *Just Being Difficult? Academic Writing in the Public Arena*

Jean-Luc Nancy, *A Finite Thinking*, edited by Simon Sparks

Theodor W. Adorno, *Can One Live After Auschwitz? A Philosophical Reader*, edited by Rolf Tiedemann

Patricia Pisters, *The Matrix of Visual Culture: Working with Deleuze in Film Theory*

Andreas Huyssen, *Present Pasts: Urban Palimpsests and the Politics of Memory*

Talal Asad, *Formations of the Secular: Christianity, Islam, Modernity*

Dorothea von Mücke, *The Rise of the Fantastic Tale*

Marc Redfield, *The Politics of Aesthetics: Nationalism, Gender, Romanticism*

Emmanuel Levinas, *On Escape*

Dan Zahavi, *Husserl's Phenomenology*

Rodolphe Gasché, *The Idea of Form: Rethinking Kant's Aesthetics*

Michael Naas, *Taking on the Tradition: Jacques Derrida and the Legacies of Deconstruction*

Herlinde Pauer-Studer, ed., *Constructions of Practical Reason: Interviews on Moral and Political Philosophy*

Jean-Luc Marion, *Being Given That: Toward a Phenomenology of Givenness*

Theodor W. Adorno and Max Horkheimer, *Dialectic of Enlightenment*

Ian Balfour, *The Rhetoric of Romantic Prophecy*

Martin Stokhof, *World and Life as One: Ethics and Ontology in Wittgenstein's Early Thought*

Gianni Vattimo, *Nietzsche: An Introduction*

Jacques Derrida, *Negotiations: Interventions and Interviews, 1971–1998*, edited by Elizabeth Rottenberg

Brett Levinson, *The Ends of Literature: The Latin American "Boom" in the Neoliberal Marketplace*

Timothy J. Reiss, *Against Autonomy: Cultural Instruments, Mutualities, and the Fictive Imagination*

Hent de Vries and Samuel Weber, eds., *Religion and Media*

Niklas Luhmann, *Theories of Distinction: Re-Describing the Descriptions of Modernity*, edited and with an introduction by William Rasch

Johannes Fabian, *Anthropology with an Attitude: Critical Essays*

Michel Henry, *I Am the Truth: Toward a Philosophy of Christianity*

Gil Anidjar, *"Our Place in Al-Andalus": Kabbalah, Philosophy, Literature in Arab-Jewish Letters*

Hélène Cixous and Jacques Derrida, *Veils*

F. R. Ankersmit, *Historical Representation*

F. R. Ankersmit, *Political Representation*

Elissa Marder, *Dead Time: Temporal Disorders in the Wake of Modernity (Baudelaire and Flaubert)*

Reinhart Koselleck, *The Practice of Conceptual History: Timing History, Spacing Concepts*

Niklas Luhmann, *The Reality of the Mass Media*

Hubert Damisch, *A Theory of /Cloud/: Toward a History of Painting*

Jean-Luc Nancy, *The Speculative Remark: (One of Hegel's bon mots)*

Jean-François Lyotard, *Soundproof Room: Malraux's Anti-Aesthetics*

Jan Patočka, *Plato and Europe*

Hubert Damisch, *Skyline: The Narcissistic City*

Isabel Hoving, *In Praise of New Travelers: Reading Caribbean Migrant Women Writers*

Richard Rand, ed., *Futures: Of Jacques Derrida*

William Rasch, *Niklas Luhmann's Modernity: The Paradoxes of Differentiation*

Jacques Derrida and Anne Dufourmantelle, *Of Hospitality*

Jean-François Lyotard, *The Confession of Augustine*

Kaja Silverman, *World Spectators*

Samuel Weber, *Institution and Interpretation: Expanded Edition*

Jeffrey S. Librett, *The Rhetoric of Cultural Dialogue: Jews and Germans in the Epoch of Emancipation*

Ulrich Baer, *Remnants of Song: Trauma and the Experience of Modernity in Charles Baudelaire and Paul Celan*

Samuel C. Wheeler III, *Deconstruction as Analytic Philosophy*

David S. Ferris, *Silent Urns: Romanticism, Hellenism, Modernity*

Rodolphe Gasché, *Of Minimal Things: Studies on the Notion of Relation*

Sarah Winter, *Freud and the Institution of Psychoanalytic Knowledge*

Samuel Weber, *The Legend of Freud: Expanded Edition*

Aris Fioretos, ed., *The Solid Letter: Readings of Friedrich Hölderlin*

J. Hillis Miller / Manuel Asensi, *Black Holes / J. Hillis Miller; or,*
  Boustrophedonic Reading
Miryam Sas, *Fault Lines: Cultural Memory and Japanese Surrealism*
Peter Schwenger, *Fantasm and Fiction: On Textual Envisioning*
Didier Maleuvre, *Museum Memories: History, Technology, Art*
Jacques Derrida, *Monolingualism of the Other; or, The Prosthesis of Origin*
Andrew Baruch Wachtel, *Making a Nation, Breaking a Nation: Literature and*
  Cultural Politics in Yugoslavia
Niklas Luhmann, *Love as Passion: The Codification of Intimacy*
Mieke Bal, ed., *The Practice of Cultural Analysis: Exposing Interdisciplinary*
  Interpretation
Jacques Derrida and Gianni Vattimo, eds., *Religion*